Sustainable Urban Regeneration

This book provides a deep insight into urban regeneration schemes and explores the parameters of what is deemed a sustainable development, before appraising existing schemes' evaluation models for the sustainable return on investment. The authors present a new practical evaluation tool that suggests quantifiable benefits for all urban regeneration stakeholders.

This new method enables the gauging of the full sustainable impact, from a given outlay of money invested in a housing-led urban regeneration scheme, through an evidence-based proof and can be used to:

- Better fulfil sustainability criteria in terms of all three aspects of the triple bottom line and contribute in a more sustainable way to address the United Nations' Sustainable Development Goal 11
- Reduce financial waste and plug the gap created by the recent economic shortfall which is impacting on housing associations, tenants and communities alike
- Evaluate historical housing-led urban regeneration schemes and model future schemes.

The method can be used as a strategic decision making or management tool, with schemes being able to be planned in, prioritised or carried out in a targeted and strategic manner; and it can be used for modelling purposes, for publicity purposes and alongside existing tools. This book provides a unique method of fully and sustainably evaluating housing-led urban regeneration schemes, useful for planners, strategic management, local authorities, housing associations, the construction industry and built environment students alike.

Dr Kevin Dean is an expert in the holistic sustainable evaluation of housing-led urban regeneration schemes and is passionate about how this knowledge can contribute in a more sustainable way to the United Nations' SDG 11.

Dr Claudia Trillo is an architect and urban planner with 25-years' experience achieved by working both in academia and in practice in several European and non-European countries. She is Reader in Architecture at the University of

Salford, UK. Her specialism lies at the interplay between architecture, urban design and regional development, with an emphasis on sustainability and Agenda 2030. She has extensively published on housing and sustainability, including two chapters on mixed income housing and housing policies for the Encyclopedia of the UN Sustainable Development Goals and a recent paper in *Housing Studies* on inclusionary housing and slums regeneration.

Professor Angela Lee's research interest lies in performance management, process management and design management within the built environment domain. She has published over 300 journal and conference publications based on her research, as well as 4 books, and managed numerous externally funded research projects to successful completion. She is presently Associate Dean in the School of Science, Engineering and Environment at the University of Salford, UK.

Sustainable Urban Regeneration

Insights and Evaluation from a UK Housing Association

Kevin Dean, Claudia Trillo and Angela Lee

Routledge
Taylor & Francis Group
LONDON AND NEW YORK

Cover image: Claudia Trillo

First published 2022
by Routledge
4 Park Square, Milton Park, Abingdon, Oxon OX14 4RN

and by Routledge
605 Third Avenue, New York, NY 10158

Routledge is an imprint of the Taylor & Francis Group, an informa business

British Library Cataloguing-in-Publication Data
A catalogue record for this book is available from the British Library

Library of Congress Cataloging-in-Publication Data
Names: Dean, Kevin (Urbanologist) author. | Trillo, Claudia, author. | Lee, Angela, author.
Title: Sustainable urban regeneration: insights and evaluation from a UK housing association / Kevin Dean, Claudia Trillo and Angela Lee.
Description: New York, NY: Routledge, 2022. | Includes bibliographical references and index.
Identifiers: LCCN 2021038807 (print) |LCCN 2021038808 (ebook) | ISBN 9780367490010 (hardback) | ISBN 9780367490003 (paperback) | ISBN 9781003043881 (ebook)
Subjects: LCSH: Sustainable urban development—Great Britain. | Urban renewal—Great Britain. | City planning—Great Britain. | Dwellings—Social aspects—Great Britain.
Classification: LCC HT169.G7 D43 2022 (print) | LCC HT169.G7 (ebook) | DDC 307.3/4160941—dc23
LC record available at https://lccn.loc.gov/2021038807
LC ebook record available at https://lccn.loc.gov/2021038808

ISBN: 978-0-367-49001-0 (hbk)
ISBN: 978-0-367-49000-3 (pbk)
ISBN: 978-1-003-04388-1 (ebk)

DOI: 10.1201/9781003043881

Typeset in Goudy
by codeMantra

Contents

List of abbreviations

ASB	Anti-Social Behaviour
BHPS	British Household Panel Survey
CEA	Cambridge Economic Associates
CLG	Communities and Local Government
CLIP	Collaboration/Conflict, legitimacy, influence and power
DEFRA	Department for Environment, Food and Rural Affairs
DETR	Department of the Environment, Transport and the Regions
DoE	Department of the Environment
EGRUP	Evaluation of Regeneration Projects and Programmes
ESA	Ecosystem Services Analysis
ESVD	Ecosystem Services Valuation Database
EZ	Enterprise Zone
FAO	Food and Agriculture Organisation
GCSE	General Certificate of Secondary Education
GKP	Global Knowledge Partnership
GP	General Practitioner
GVE	Global Value Exchange
HACT	Housing Associations' Charitable Trust
HCA	Homes and Communities Agency
HMSO	Her Majesty's Stationery Office
IUCN	International Union for Conservation of Nature
LGA	Local Government Association
LIBOR	London Interbank Offered Rate
LUDA	Improving the quality of life in large urban distressed areas
MSP	Multi-Stakeholder Partnership
NDC	New Deal for Communities
NEF	New Economics Foundation
NHS	National Health Service
NICE	National Institute for Health and Care Excellence
NPC	New Philanthropy Capital
ODPM	Office of the Deputy Prime Minster
OED	Oxford English Dictionary
ONS	Office for National Statistics

REDF	Roberts Enterprise Development Fund
RICS	The Royal Institution of Chartered Surveyors
ROAMEF	Rationale, Objectives, Appraisal, Monitoring, Evaluation and Feedback
SCBA	Social cost benefit analysis
SDG	Sustainable Development Goal
SRB	Single Regeneration Budget
SROI	Social Return on Investment
SuROI	Sustainable Return on Investment
SURPAM	The Sustainable Urban Renewal Project Assessment Model
TEEB	The Economics of Ecosystems and Biodiversity
UDC	Urban Development Corporation
UDC	Urban Development Corporations
UK	United Kingdom
UN	United Nations
UNCED	United Nations Conference on Environment and Development
UNEP	United Nations Environment Programme
UNESCO	United Nations Educational, Scientific and Cultural Organisation
USA	United States of America
WCED	World Commission on Environment and Development
WHO	World Health Organisation
WWF	World Wildlife Fund

Foreword

Among the many volumes devoted to urban regeneration and housing, little is mentioned of the subject that forms the central concern of this book: the important topic of how to ascertain the sustainable social return on investment.

In many sense the ideas explored herein represent the practical application of the central principles of sustainable development, especially the need to balance the social, environmental and economic components in order to deliver a triple bottom line. Through developing and applying the social return on investment model, the authors are able to make explicit what are often obscure or difficult choices. By identifying and assessing such 'trade offs', the authors are able through their SuHousingImpact tool to demonstrate the full range of costs and benefits associated with many projects.

This book explores many aspects of the various dilemmas that frequently confront housing professionals and other decision makers and, in doing so, it creates a means of making difficult choices through the monetisation of the effects of development schemes. However, as the authors argue, this approach should not be used in a slavish manner, but rather should be employed in order to make explicit the choices that are inherent in many development schemes. Equally important, the authors acknowledge that housing development choices cannot be dictated by financial issues alone; indeed, they argue that a full sustainable development approach must be adopted.

Most importantly, the model which forms the central concerns of the book offers a means of understanding the long-term consequences of investment in more energy efficient housing. This alone is a very valuable contribution to the literature.

This is an important book that deals with an underdeveloped subject, and it does this in a highly readable and practical manner. The authors shine a light on what is often considered to be an obscure aspect of housing investment, and they are to be congratulated for the contribution that they make.

Professor Peter Roberts
Chair of the Northern Ireland Housing Executive
Professor Emeritus, University of Leeds
Co-Chair of the NHS Confederation Sustainable Development Group
Honorary Professor at Queen's University Belfast
Visiting Professor at the University of Ulster

1 Introduction

The UN Urban Agenda, economic climate and solutions for housing associations

Introduction

Urban regeneration has a substantial impact to play on all three dimensions of sustainability –often referred to as the three pillars: society, economics and the environment – or more commonly termed as the *triple bottom line* (Elkington et al., 2007). It is therefore an activity of considerable importance in achieving a more sustainable society. The UK Government has integrated the goal of sustainability into urban regeneration policies, yet the proliferation of definitions and conceptualisations of sustainability render the term to be so poorly understood and slippery that it can be easily misinterpreted, or at worse, even manipulated.

The new 2016 United Nations' (UN) Urban Agenda reaffirms the concept that sustainable cities require both environmental and social sustainability, and with this introduced 17 [interconnected] Sustainable Development Goals (SDGs) that support a better and sustainable future for all by addressing issues relating to poverty, inequality, climate change, environmental degradation, peace and justice (see https://www.un.org/sustainabledevelopment/sustainable-development-goals/). With goal 11 (SDG11) – that of making cities inclusive, safe, resilient and sustainable – a target of sufficient and affordable housing has been set (https://sustainabledevelopment.un.org/sdg11). Despite housing's central role and importance in ensuring sustainability (Hills, 2007), and despite the importance of both the environmental and social pillars of the triple bottom line in ensuring sustainable development (Elkington et al., 2007), current evaluative methods that support decision making on social housing interventions still fail in capturing all of the socio-environmental spillovers of the UN's SDG11. This book contributes to addressing this issue by providing decision makers with a new evaluative tool, namely 'SuHousingImpact', building on and taking forward the Sustainable Return on Investment (SuROI) approach (Bichard, 2015), and allowing the successful capturing of such socio-environmental spillovers in housing-led regeneration schemes.

The book introduces and covers the subject domains of urban regeneration, with a focus on housing-led regeneration, together with sustainable development and evaluation. In addition to this, existing methods of evaluation of housing-led urban regeneration schemes are critiqued. The approach of SuROI (Bichard, 2015)

DOI: 10.1201/9781003043881-1

is introduced as an extension of the Social Return on Investment (SROI) methodology, which has been used within the context of the built environment previously (Aspden et al., 2012; Bichard, 2015; Bridgeman et al., 2015, 2016; Watson et al., 2016; Watson and Whitley, 2016).

The research in this book links in with the concept of "*social innovation*". Social innovation is a domain which is becoming "increasingly evident in policy [and] scientific and public debates" (Howaldt et al., 2014) and is considered as being "increasingly influential in both scholarship and policy" (Moulaert et al., 2013), with a growing consensus emerging amongst practitioners that "widespread social innovation is required to cope with the significant challenges that societies are facing [both] now and in the future" (Howaldt et al., 2014). The concept has been referred to as "a new combination of social practices in certain areas of action or social contexts with the goal of better satisfying or answering social needs and problems than is possible on the basis of existing practices" (Ibid, 2014). Specifically, this involves ideas which have been subsequently turned into practical approaches (Evers et al., 2014).

Social innovation approaches have been said to be important for finding solutions to the interconnected challenges of global and sustainable development (Babu and Pinstrup-Andersen, 2009), whilst the significance of methods and technologies has been widely acknowledged to be central to innovation studies (Howaldt et al., 2014).

Such innovation is outlined in this book, which consequently leads to the creation and unveiling of the novel 'SuHousingImpact' tool, useful to sustainably evaluate housing-led urban regeneration schemes in both an ex-ante and ex-post context. This book demonstrates the implementation of the tool on two real historical case studies from a leading housing association based in the North West of England, UK, through two extant housing-led regeneration schemes – a not-for-profit housing association; those schemes being an environmental-led scheme and a high-rise scheme, with results showing that the environmental and social spillovers are largely disregarded because of a gap in the evaluation methods and that room for significant improvements exists. This proves the importance of the 'SuHousingImpact' tool, which allows the unveiling of hidden social and environmental benefits from housing-led urban regeneration schemes and therefore supports a better alignment of current practices to the pursuing of the SDG11. It is hoped that this new tool can be used to better inform and evaluate when compared to previous methods of evaluation in urban regeneration.

Background and context

Economic climate and consequences for housing associations

Housing associations are currently operating under severe financial constraints, which make the decision making process extremely demanding, since every choice needs to be thoroughly assessed in terms of benefits. This situation has become particularly harsh in recent times, while in the past less scarcity of resources allowed more flexibility in choices. Before the economic crash of 2008 in the United Kingdom, housing associations could rely on significant bank financing to fund their development. Banks and other lenders historically provided housing associations with loans in favourable terms. Until the advent of the credit crunch, banks priced loans to housing associations at 20–30 basis points above the London interbank lending rate (House of Commons, Communities and Local Government Committee, 2009). However, once the credit crunch came about, not only did banks charge more for new loans, but they also sought to rewrite the already existing housing association loans they had outstanding. Those housing associations seeking additional financing found that banks demanded rates as high as 300 basis points above LIBOR (London Interbank Offered Rate), often more than ten times higher than their previous rates. A number of lenders refused altogether to lend to housing associations (Hilditch, 2009; House of Commons, Communities and Local Government Committee, 2009). Banks also sharply increased the cost of credit for housing associations, whilst demand for for-sale homes and shared-ownership housing plummeted, reducing revenue for housing associations (Dowler, 2009). More recently as part of the housing white paper, the UK Government has announced that rent decreases (explained further below) are to remain in place until 2020 and that there remains a focus on Right to Buy (UK Government, 2017).

Housing associations had increasingly been building housing for sale in the open market to generate additional revenue to help fund the development of social housing. However, the financial crisis in 2008 left housing associations with thousands of unsold housing units, whilst additionally reducing the market value of the associations' housing stock and land holdings. Additionally, housing associations' ability to generate funds to support the development of social housing was curtailed, whilst the crisis also reduced the amount of housing acquired from private developers through Section 106 agreements (planning obligations under Section 106 of the UK's Town and Country Planning Act 1990 is a mechanism which makes a development proposal acceptable in planning terms, that would not otherwise be acceptable). A reduced revenue subsidy for housing, notably via housing benefit, has contributed to the financial difficulties that housing associations currently face. In addition, a reduction of welfare support will impact on the ability of tenants to pay their rent, whilst the cutting of rent paid by tenants by 1% has added to the challenges faced. To cap it all off, the Right to Buy scheme can force associations to sell at huge discounts. Added to this, falling wages and a lack of employment contributes to these issues, all this at a time when social housing is likely to be in even greater demand (Chevin, 2013), putting even more pressure on housing associations. Consequently, housing associations have had to be more economically aware. Poor decision making can have major consequences. It is therefore important to make sure that investment is used efficiently, and that the amounts of money which are available are not wasted.

Presently, the economic climate is now even more demanding; additionally, with the added impact of the recent world pandemic of Covid-19, it is increasingly important that the evaluation of housing-led urban regeneration schemes is of a high standard, and that the conclusions coming about from a particular evaluation are accurate. To this end, it is imperative that all benefits arising from a particular scheme are taken into consideration, and that as much information as possible on the scheme impacts is available for strategic decision makers. However, there is an absence of frameworks available to assess impacts in terms of sustainable development on the built environment (Thomson et al., 2009), or certainly a limited number of such frameworks (DETR, 1998; OECD, 2000); and the absence of appropriate frameworks has often been considered as playing a part in the inability to deliver the desired objectives of urban regeneration schemes (Kazmierczak et al., 2009) which can have negative repercussions, especially when taking into consideration the aforementioned current economic climate, when money is not always readily available and mistakes are potentially costly. Along these lines, Tyler et al. (2013:171) pose the question as to why, despite the resources that many countries commit to urban regeneration, has there been "so little evidence available on the aggregate value of regeneration benefits?"

Economic climate, its effect on housing associations, the resulting problem and potential solution

A housing association's core role is to provide housing for those who are in greatest need. The sector in the United Kingdom contains the largest number of professionally managed properties (approximately 3.9 million, 17% of all households; MHLC, 2019), and as such, has a significant role to play in improving the country's economic, environmental and social sustainability. Improving the sustainability of existing housing stock is a major challenge facing the UK social housing sector, for which there is little support to navigate the growing and often incongruent information relating to sustainable development and how to operationalise it.

The housing association sector has typically been underpinned in previous years by high levels of capital subsidy from the UK Government (Malpass, 2005), with the majority of revenue being funded indirectly through Housing Benefit (Steele, 2012). In recent times, there has been a change from local councils, to now housing associations developing most rental housing, initially receiving substantial government funding to do so (Schwartz, 2011). Nowadays there involves a greater reliance on private actors, market mechanisms and commercial capital, with the shift being described as a "migration from the public sector towards the private market" (Blessing, 2015:198).

However, despite this new commercial direction, things have become more difficult for housing associations in recent years since the economic crash of 2008 and the economic catastrophe that is Covid-19. According to Evans et al. (2016), the financial crisis "affected the housing sector the hardest".

Specific areas have consequently contributed to the necessity for change in the way housing associations do business, resulting in a more commercial approach being taken. Firstly, a reduced revenue subsidy for housing, notably via housing benefit, has contributed to the financial difficulties that housing associations currently face. In addition, a reduction of welfare support will impact on the ability of tenants to pay their rent. Added to this, falling wages and a lack of employment add to the issues, whilst social housing is likely to be in even greater demand (Chevin, 2013), putting even more pressure on housing associations.

The change to welfare support is another driver for change. In late 2010, the UK Government began a radical overhaul of almost all aspects of welfare support. Central to these reforms was the government's aim to reduce the nation's benefits bill, particularly in relation to housing benefit. Because housing benefit is no longer paid direct to the housing association via direct debit, this can affect their rental income. In addition, the so-called 'bedroom tax' (in 2016, the UK Government introduced an under-occupancy penalty/charge from the Welfare Reform Act 2012 whereby tenants living in social housing with rooms deemed 'spare' faced a reduction in Housing Benefit, resulting in them being obliged to fund this reduction from their own incomes or face rent arrears and potential eviction) has added to the challenges faced, whilst the Right to Buy scheme can force associations to sell at huge discounts.

Bearing the economic effects of the credit crunch in mind, housing associations have tended to react by attempting to plug monetary gaps via a drive to be more commercial. Chevin (2013:7) writes that many housing associations were looking at accomplishing "greater value for money and better asset management", together with looking into the possibility of additional income streams. Housing associations will ultimately have to explore new funding mechanisms and adopt more diverse business models to avoid becoming marginalised as government funding becomes increasingly constrained (Places for People, 2017).

Indeed, there have been reports of more diversified activities providing up to £3.5 billion of housing turnover (Social Housing, 2016), and additionally a sharp rise in housing association turnover from commercial activity has been highlighted (Inside Housing, 2014).

Other ways forward in the future to supplement housing associations' income include diversifying, widening and strengthening services provided, including potential partnerships with National Health Service (NHS) Trusts and GPs (general practitioners) to provide outreach health care, working with local authorities to re-design local care services around the home to maximise independence and well-being for people who need care, providing specialist support services to vulnerable people with multiple or complex needs, delivering government programmes for supporting people without jobs into work, investing in schools, helping improve educational achievement, employment prospects and life chances, rehabilitation and resettlement of offenders and helping to reduce reoffending and subsequently easing pressures on prisons. An emphasis will be on a private business that generates profits to utilise for social good – a "Profit for Purpose" business (NHF, 2014).

The new stance that housing associations have taken commercially highlights some areas to be mindful of. For example, the relationship between the housing association and their tenants could be compromised if value for money on rent levels and arrears is sought; there is a concern amongst some observers that community activities and the sector's historically social ethos are increasingly taking a back seat, and there is a fear that housing associations will be indistinguishable from private sector companies or that as associations seek greater efficiencies, their investment and involvement in community activities will be reduced (Chevin, 2013).

Housing associations are therefore typically finding it more difficult than in previous times by virtue of the lack of public money available. Consequently, housing associations have had to be more economically aware. Poor decision making can have major consequences. It is therefore important to make sure that investment is used efficiently and that the amounts of money which are available are not wasted.

Summary

In this chapter, an introduction to the book was provided, with the background and context being covered. This included an introduction to urban regeneration and sustainability and the notion of the *triple bottom line*. The current economic climate and its consequences for housing associations was discussed and outlined, together with the resulting problem and potential solutions. Chapter 2 progresses onto urban regeneration and sustainable development.

2 Urban regeneration and sustainable development

Introduction

This chapter demystifies the subject domains of sustainable development and urban regeneration and additionally covers the history and definitions of urban regeneration. It then proceeds to look at the links between housing and urban regeneration, housing and health, as well as housing and its relationship with each part of the triple bottom line (Elkington et al., 2007). The economic climate and its effect on housing associations is looked into in more depth, as are the corresponding problems and potential solutions. The importance of housing policies and the negative impact the new economic climate could determine is considered within the broader framework of the Sustainable Development Goals. At this point, purpose, history and definitions of sustainable development are made clear, together with definitions of 'weak' versus 'strong' sustainability, models of sustainable development, 'capitals', the recent UN Habitat III conference and its shift towards sustainable cities, and finally, sustainable development and its links with urban regeneration.

Urban regeneration

The population of those living in urban areas is increasing. The United Nations (2018) states that around 55% of the world's population is living in urban areas and predicts that by 2050, this number will increase to 68% – similar to the UN's earlier predictions published in 2012. Indeed, the United Nations (2012a) additionally predicts that the land conversion to urban areas will triple by 2030 and asserts that 21st-century global urban trends have "taken human civilisation into unknown territory" with half the world's population living in urban areas for the first time in history, stating that by 2030, the world urban population is expected to be around 5 billion people, whilst the level of urbanisation will rise from what was 52% in 2011 to a figure of 67% in 2050 (UN, 2012b; UN Habitat, 2011). Evans et al. (2016) supports this stance and predicts a bigger increase: that three quarters of the world's population will live in urban areas by 2050. Despite variations in predictions, all sources agree that the numbers are on the increase, and by substantial amounts.

DOI: 10.1201/9781003043881-2

In Western countries, in order to accommodate the shifts referred to above, development through urban regeneration schemes needs to be undertaken (Rydin et al., 2012).

Urban regeneration can consequently be considered as an important activity, and this importance has been discussed widely. It has been referred to as one of the most fundamental aspects of social intervention that a government can implicate itself in (Diamond et al., 2006). In the United Kingdom context, it has been stated that without regeneration, the most deprived communities in the United Kingdom would have little chance of economic recovery (The Work Foundation, 2012), and additionally, its long-term impacts within society are highlighted (Yau and Chan, 2008). Tallon (2009) and Roult and Lefebvre (2013) note the rise of urban regeneration, citing that there has been a shift towards the idea of urban regeneration gradually taking place for about four decades.

Urban regeneration has been referred to as an activity that attempts to address urban problems via economic, social and environmental improvements in a particular area (Roberts and Sykes, 2000). The reference to the triple bottom line of economic, social and environmental aspects is commonplace within the literature. Authors such as Roberts and Sykes (2000), Ercan (2011), Brown (2006), Lombardi et al. (2011), Scottish Government (2011), Smith (2006), Brandon and Lombardi (2011) and Ha (2007) all refer to the concept of the triple bottom line as being integral to urban regeneration.

The current stance differs from the historical one. Historically, it is claimed that urban regeneration was carried out with only physical notions in mind (Atkinson and Moon, 1994; Couch et al., 2011) or alternatively that there was a focus on economic and environmental aspects only (Colantonio and Dixon, 2009), thus omitting the social component. Additionally, the aim appeared to have previously been one of involving the achievement of quantity rather than quality, especially after the destruction caused as a result of the Second World War (Priemus et al., 1999).

Definitions

There are many views and opinions within the literature on what exactly constitutes the definition of 'urban regeneration'. If we firstly look at the dictionary definition of the two words *urban* and *regeneration*, the definitions appear as follows (OED, 2017):

Urban *adjective*
In, relating to, or characteristic of a town or city

Regeneration *noun*
The action or process of regenerating or being regenerated
From "re-generare" – to create again

Regenerate *verb*
Bring new and more vigorous life to (an area etc.); to revive, especially in economic terms

Thus, 'urban' means "relating to towns or cities" (Tallon, 2009). According to the OED (2017), the term 'regeneration' is used when any revival of an area is being undertaken, whilst Turok (2005) uses the term to describe the reinvigorating of areas, Williams (2002, 2007) mentions the term 'reborn' and Clements (2005) adds to this with the term 'renewal'.

This 'renewal' can mean the physical environment, including: the aesthetics of an area, the economic environment or the social environment. Indeed, Tallon (2009) states that regeneration "has come to be associated with any development that is taking place in (UK) towns or cities". Roberts and Sykes (2000) define the term 'urban regeneration' as

> a comprehensive and integrated vision and action which leads to the resolution of urban problems and which seeks to bring about a lasting improvement in the economic, physical, social and environmental conditions of an area that has been subject to change.

However, Turok (2005) disputes, stating that regeneration is rarely classed as being comprehensive. Further, Gibson and Kocabas (2001) state that urban regeneration is a holistic, comprehensive and integrated approach that should embrace the three key features of economy, equity and environment. The economy and the environment contribute to part of the UK Government's definition of the term. In the 2011 Governmental publication *Regeneration to enable growth*, it is defined as "the broad process of reversing physical, economic and social decline in an area where market forces will not do this without intervention" (DCLG, 2011). This can be differentiated from Jones and Evans' view that regeneration is considered as being "the large scale process of adapting the existing built environment with varying degrees of direction from the state" (Jones and Evans, 2008). Jones and Evans (2008) also state that regeneration is a type of intervention particularly well suited to de-industrialised cities that are attempting to join the new and changing economy, thus requiring public investment to change their physical infrastructure to adapt to this changing economy. The Audit Commission asserts that the general aim of regeneration is "to enable communities that have suffered from economic, social and environmental decline to 'work' again" (Audit Commission, 2003). However, Williams (2002) highlights only one factor, citing that the single issue is never poverty, racism or social inclusion, but the environment. Clements (2005) states that at its simplest level, urban regeneration implies the renewal of lost *vitality*, and again refers to the triple bottom line, adding… "whether physical, social or economic (or some combination thereof)", whilst alternatively, Williams suggests the term *reborn* (Williams, 2007).

Turok (2005) cites that three typical features of urban regeneration involve the changing of the "nature of a place" as well as the processes and persons involved in this change, and that there are often "multiple objectives and activities" that go beyond what a central government would readily accept as being in their jurisdiction and that partnership working is a necessity of any process. He also states that there are other benefits of urban regeneration which include three basic concepts revolving around:

a Improving the lives, skills and aspirations of the persons living in an area;
b The attempt to reinvigorate an area in terms of business performance which would in turn make an area more economically prosperous; and
c To make an area more "generally appealing" in order to facilitate a) and b).

This definition again honing in on the potentially holistic nature of urban regeneration.

According to Roberts and Sykes (2000), there are certain fundamental principles involved in any form of urban regeneration that should:

- Be based on a detailed analysis of the condition of an urban area;
- Be aimed at simultaneous adaptation of the physical fabric, social structures, economic base and environmental condition of an urban area;
- Attempt to achieve this simultaneous adaptation through the generation and implementation of a comprehensive, integrated and balanced strategy;
- Ensure that the strategy and implementation are developed in accordance with aims of sustainable development;
- Set clear operational objectives which where possible should be quantified;
- Make the best possible use of existing resources, including natural, economic and human resources "including land and existing features of the built environment";
- Seek to ensure the maximum amount of consensus through consultation and communication with all stakeholders with a legitimate interest in the regeneration of the area in question;
- Measure the progress of strategy devised against the agreed and specified objectives and should monitor "the changing nature and influence of the internal and external forces which act upon urban areas";
- Recognise the need for continual revision of programmes of implementation as changes occur;
- Recognise that different elements of a strategy may unravel at different speeds, thus requiring the redirection of resources or the provision of different resources where necessary.

The authors add that the theory and practice of urban regeneration together shares commonalities:

- It is typically an interventionalist activity.
- It is typically an activity that straddles the public, private and voluntary and community sectors.
- It is typically an activity that is likely to experience considerable changes in its institutional structures over time in response to changing economic, social, environmental and political circumstances.

- It is typically a means of mobilising collective effort and providing the basis for the negotiation of appropriate solutions.
- It is typically a means of determining policies and actions designed to improve the condition of urban areas and developing institutional structures necessary to support the preparation of specific proposals.

In addition to the above guidance, the Department for Communities and Local Government (2008) adds that urban regeneration should aim to:

- Secure long-term change, by tackling barriers to growth and reducing worklessness;
- Improve places and make them more attractive to residents and the investment community so that new and existing businesses can prosper;
- Foster ambition and unlock potential in the most deprived areas by breaking out of the cycles of poverty in an area;
- Enable everyone in society to be empowered to participate in decision making and to take advantage of the economic opportunities that regeneration brings;
- Deliver sustainable development, which contributes to people's satisfaction with where they live as well as wider government goals; and
- Open up opportunities to create more equal communities.

As can be seen from the literature, there are a wide variety of definitions and opinions on what exactly constitutes the term 'urban regeneration'. A variety of these definitions and opinions are accumulated in Table 2.1.

The definitions presented in Table 2.1 share some common traits, but also possess many differences and so don't link together perfectly. Tallon (2009) and the Oxford English Dictionary (2017) both share the view that regeneration relates to towns or cities; whereas Brown (2006) refers to neighbourhoods; the Scottish Government (2011) refers to places; the International Federation of Arts Councils and Culture Agencies (2006), ODPM (2001) and LGA (2000) refer to areas or places; whilst Ercan (2011) refers to urban areas. Along these lines, Jones and Evans (2008) mention that industrialised cities are particularly well suited to regeneration. Tallon (2009) refers to regeneration as a 'development', whereas the UK Government (2011), Scottish Government (2011) and Jones and Evans (2008) refer to a 'process', and the LGA (2000) refers to a 'transformation'. Roberts and Sykes (2000) and Brown (2006) describe urban regeneration as an 'action', whilst Ercan (2011) and Roberts and Sykes (2000) comment on a 'vision', whilst Gibson and Kocabas (2001) refer to an 'approach'. Roberts and Sykes (2000), Brown (2006), Scottish Government (2011), Ercan (2011), ODPM (2001) and LGA (2000) refer to triple bottom line factors, whilst Turok (2005) mentions the economic and social aspects and infers an improvement to the physical and environmental. Clements (2005) writes something very similar, mentioning that the necessary improvements needed involve physical, social or economic aspects. Gibson and Kocabas (2001) differ from this in that they state that the economy, equity and

Table 2.1 Definitions of urban regeneration

Author	*Definitions include*
Academic literature	
Roberts and Sykes (2000)	A comprehensive and integrated vision and action which leads to the resolution of urban problems and which seeks to bring about a lasting improvement in the economic, physical, social and environmental conditions of an area that has been subject to change
Gibson and Kocabas (2001)	Is a holistic, comprehensive and integrated approach which should embrace three key features of economy, equity and environment
Williams (2002, 2007)	The single issue is never poverty, racism or social inclusion. It's the environment; Regeneration is about areas being reborn
Clements (2005)	At its simplest level, it implies the renewal of lost vitality, whether physical, social or economic (or some combination thereof)
Turok (2005)	Rarely classed as being comprehensive; Involves changing the nature of a place, multiple objectives and activities that go beyond what central government would accept as being within their jurisdiction; Partnership working is a necessity of any process; Improves lives, skills and aspirations of persons living in an area; Attempts to reinvigorate an area in terms of business performance; Makes an area more economically prosperous; Makes an area more generally appealing
Brown (2006)	Urban regeneration is concerted social, economic and physical action to help people in neighbourhoods experiencing multiple deprivation reverse decline and create sustainable communities
Jones and Evans (2008)	The large-scale process of adapting the existing built environment with varying degrees of direction from the state; Regeneration is a type of intervention particularly well suited to deindustrialised cities that are attempting to join the new and changing economy, thus requiring public investment to change their physical infrastructure to adapt to the changing economy
Tallon (2009)	Relating to towns or cities; Associated with any development that is taking place in towns or cities
Ercan (2011)	A comprehensive integration of vision and action aimed at resolving the multi-faceted problems of deprived urban areas to improve their economic, physical, social and environmental conditions
Oxford English Dictionary (2017)	In, relating to, or characteristic of a town or city; The revival of an area
Non-academic literature	
LGA (2000)	The transformation of a place – residential, commercial or open space – that has displayed the symptoms of physical, social and/or economic decline; breathing new life and vitality into an ailing community, industry and area [bringing] sustainable, long-term improvements to local quality of life, including economic, social and environmental needs

Author	*Definitions include*
ODPM (2001)	Not simply about bricks and mortar. It's about the physical, social and economic wellbeing of an area; it's about the quality of life in our neighbourhoods. In relation to the physical, this is as much about the quality of public realm as it is about the buildings themselves
Audit Commission National Study (2003)	The aim of regeneration in general is to enable communities that have suffered from economic, social and environmental decline to 'work' again
International Federation of Arts Councils and Culture Agencies (2006)	The renewal, revival, revitalisation or transformation of a place or community. It is a response to decline or degeneration. Regeneration is both a process and an outcome
Scottish Government (2011)	The holistic process of reversing the economic, physical and social decline of places where market forces won't suffice
UK Government (2011)	The broad process of reversing physical, economic and social decline in an area where market forces will not do this without intervention

the environment are the three key features involved in regeneration, with Williams (2002, 2007) stating that the environment is the key feature involved. The UK Government (2011) refers to the necessary reversal of the three features of physical, economic and social decline and the Audit Commission National Study (2003) echoes this, by adding that it is economic, social and environmental decline that creates the need for urban regeneration. Jones and Evans (2008) mention a need for deindustrialised cities to change their physical infrastructure.

Jones and Evans (2008) go on to describe the process as one of adaptation, whereas Turok (2005) mentions the "changing of the nature" of a place. Lastly, Turok (2005) and Gibson and Kocabas (2001) both agree that the process is a holistic one, with Turok (2005) mentioning the multiple objectives involved in the process. Roberts and Sykes (2000) and Gibson and Kocabas (2001) disagree with Turok (2005) in terms of whether urban regeneration is comprehensive or not, with the former authors stating that it is and the latter stating that it isn't. Thus, with competing definitions, it is apparent that misunderstanding and misinterpretation can easily ensue.

History of urban regeneration in the United Kingdom

Urban regeneration in the United Kingdom has evolved over time. The 1950s were preoccupied with post war reconstruction and was considered as a period of dramatic suburban growth in new UK council housing estates. Relocation of industry resulted in severe decline of inner city areas (Tsenkova, 2003), whilst towns and cities were based primarily on master plans and suburban growth, and at that time, received only minimal intervention from the private sector (Roberts and Sykes, 2000).

The 1960s continued with much of what was utilised in the 1950s, with the 1950s theme continuing (Tsenkova, 2003). The State planning system was considered as incapable and inefficient (Barnekov et al., 1989), but despite this, urban regeneration was still typically public sector driven and was concerned primarily with the large-scale development of inner city slum areas (Couch, 1990).

The 1970s focussed on in situ renewal and neighbourhood schemes, together with suburban development. There was also a "shift in emphasis" regarding the physical or "bricks and mortar" approach to regeneration (Tsenkova, 2003). Harding and Garside (1995) state that in the late 1970s there was a focus on the "economic regeneration of selected areas", such as large-scale infrastructure redevelopment, fuelled by increased private sector involvement – such as can be seen with the London Docklands. Indeed, Thornley (1993) adds that the advent of the UK Conservative Government in 1979 brought the private sector to the fore within the field of urban regeneration. Barnekov et al. (1989) adds that urban policy within this decade focussed on four main areas:

- Rising urban poverty, housing needs, low income earners and unemployment
- Long-term unemployment and increasing job loss in inner city areas
- Concentrations of racial minorities in major urban centres
- The causes of decline – as opposed to the symptoms

In the 1980s, major schemes of development and redevelopment, flagship projects and out of town projects were conceived. The "Regeneration Blueprint" of the era was the 1977 White Paper *Policy for the Inner Cities* (Department of the Environment, 1977) which concluded that it was the decline of economic establishments that was responsible for the deterioration of the inner city (Deakin and Edwards, 1993). There was a realisation that urban regeneration could not be carried out to the levels needed, without the notion of partnerships and partnership working (Tsenkova, 2003).

The 1980s also hailed the arrival of two broad initiatives which "accelerated the process of urban regeneration" and provided the vehicles to create such partnerships (Ibid, 2003):

- Urban Development Corporations (UDCs)

The remit of UDCs was to bring buildings and land back into effective use, to develop industry and to create an attractive environment for people to work and live in (Berry et al., 1993). UDCs were brought about to strategically combat previous public sector driven policies and were classed as "enablers". By this, it is meant that they were not "bound by local authority plans" (Berry et al., 1993; Brownill, 1990) and that certain powers were held by UDCs regarding such aspects as the granting of planning permission and/or compulsory purchase orders (Ibid, 1993).

UDCs also had the power to grant financial aid and were responsible for any infrastructure within a designated area (Tsenkova, 2003).

- Enterprise Zones

Enterprise Zones were designated areas free from planning regulations. The zones were designed to promote aesthetic creativity (Berry et al., 1993). Enterprise Zones were also able to claim economic incentives and financial allowances (Lawless, 1989).

The view of the time, a view that had changed from previous decades, was that regeneration was best carried out by the private sector and that local authorities would be "relegated to a minor role" (Deakin and Edwards, 1993).

In the 1990s, a more sustainable, partnership-based and integrated approach was adopted with emphasis on the role of communities (Roberts and Sykes, 2000). Imrie and Raco (2003) disagreed with this, stating that in the 1980s and 1990s, the approach to regeneration was largely property-led. The holistic approach emphasised in the Urban Task Force report, produced by the Department of the Environment, Transport and the Regions in 1999 (Urban Task Force, 1999), stressed that urban regeneration should be founded on strong democratic local leadership, public participation and the use of public finance to attract increased private investment. Local and National Government, private industry and local communities began working together in a more comprehensive approach to regeneration, through social, economic and environmental policies (Tsenkova, 2003).

In the present time, urban regeneration attempts to impact upon such aspects as the economy, employment, economic competitiveness, social exclusion, community issues, vacant or deteriorated sites in cities, land and property requirements, environmental quality and sustainable development (Keles, 2003; Roberts and Sykes, 2000; Turok, 2005).

Housing-led urban regeneration

The field of urban regeneration incorporates many different domains, including those such as retail-led regeneration and property-led regeneration (Tallon, 2009). The field of housing-led regeneration is the focus of this book.

Access to housing is a human right (Universal Declaration of Human Rights, 1948; Article 25 recognises the right to housing as part of the right to an adequate standard of living) which has a multidimensional impact on human life (UN, 2016). Because housing has been referred to as possessing "social, environmental and economic dimensions which are all closely interrelated", it is cited as being "much more than providing people with a place to live" (UN, 2016).

Housing can be said to play a 'pivotal role' in the regeneration processes being carried out in urban areas (Ha, 2007:118) and can also be classed as a driving force of regeneration schemes (Haran et al., 2011) or as being vitally important (ODPM, 2003). Housing has been referred to as being one of the most important public policies affecting urban development and consequently has significant potential to contribute to sustainable development (Priemus and ten Heuvelhof, 2005; Tosics, 2004). It is often seen as an indicator for growth and sustainable development of an area (Winston, 2009) and is considered to be at the centre of sustainable regeneration policy (Maliene and Malys, 2009). Thus, as it plays such a pivotal role in the overall regeneration processes in urban areas (Ha, 2007), it is especially important to evaluate.

Bailey (2010) described the housing sector as a symbol that represents the entire scope of urban development and regeneration process, which should be considered as the "heartbeat" of the regeneration concept that is able to not only drive local regeneration but which can also provide substantial benefits in terms of creating sustainable local communities (Smith, 2006).

Housing is additionally important on a day-to-day level for many people. Hornberg and Pauli (2011:276) assert that with the average time spent indoors every day, especially at home, being up to 90% in industrialised nations, this shows the importance of housing for wellbeing and quality of life, as well as a home and retreat which "affords protection and safety from the outside world and room for personal growth". Paddison (2012) adds that housing and neighbourhood quality have profound implications for the quality of life enjoyed, or endured by residents, and Hills (2007:xii) writes that housing is vital to the economy, environment, every individual and family "for whom a home represents so much more than just a place to live".

Moving onto housing as a regeneration vehicle, it is considered that housing associations, as a major vehicle of delivering urban regeneration, are at the forefront of contemporary regeneration activity in neighbourhoods (CLES, 2015), whilst MacLennon (2007:7) states that good housing outcomes facilitate neighbourhood renewal, poverty reduction and wider economic progress and adds that "housing is central to the quality of life in our communities, to fulfilling people's aspirations, to the economy and to the environment". Maliene and Malys (2009:426) additionally add that "housing is a key issue to consider in delivering healthy and attractive communities".

Housing and health

Housing quality is an important determinant of health (McGraw, 2007). Indeed, not only is poor housing "associated with poor health", but the links between housing and health "suggest that housing improvement may well be justified on health grounds alone" (Thomson et al., 2013:4).

Indeed, Jacobs et al. (2010) cite that for more than a century, society has recognised that improved housing generally leads to improved health, with sufficient evidence now showing that specific housing interventions can improve health outcomes. The link between housing and health has been made explicit by the Public Health Green Paper (Department of Health, 1998) where it was formally acknowledged that housing was a key health determinant and aspect of inequality. Recent research shows that poor housing in the United Kingdom costs the National Health Service (NHS) £1.4 billion a year (House of Lords, 2015).

Stewart (2005:526) writes that there is a "wide recognition that decent housing can help maintain and promote health, whilst poor housing has a negative health impact". Additionally, Thomson and Thomas (2015:205) report that a number of "persistent links" have been reported between poor housing and poor health. Such links include immediate issues, such as the provision of an affordable level of thermal comfort which can improve health and prevent adverse effects of cold (Thomson and Thomas, 2015), or the likes of poor domestic conditions, cold and damp/fuel poverty, emotional health or depression or overcrowding which can lead to disease, or improvements to affordable warmth, ventilation and exposure to damp (Basham et al., 2004; Caldwell et al., 2001). In addition, it has been claimed that education can suffer due to a lack of space for children to work within an abode, whilst noise pollution, pest infestation, local crime and anti-social behaviour are all areas noted (Stewart, 2005).

Somerville et al. (2000) further this by explaining the possible longer term effects of good quality housing; that improved respiratory health following housing improvements can lead to reduced absences of children from school, which can have educational benefits further down the line, whilst Egan et al. (2013) cite further longer term worries when stating that an exposure to poor housing can increase the risk of mortality beyond what might be predicted from individual level socio-economic factors. It is increasingly being recognised that the wider housing environment location represents access to employment, training, facilities, decent food and social cohesion (Stewart and Rhoden, 2006).

In addition to the effect of housing on physical health, Johnson (2013:32) adds that housing has a "significant impact on mental health". The growing body of evidence concerned with emotional/mental ill health attributable to poor housing suggests that the experience can lead to delayed communication, disruption to education and can contribute to behavioural and mental health problems (Ballinger, 2002; Bogard et al., 1999), whilst the positive effects of housing improvement on mental health have been "consistently reported" (Thomson et al., 2007:211).

The general wellbeing of an individual is also affected by housing. Bond et al. (2012:1) write that it is not only the quality and aesthetics of housing which are associated with wellbeing, but feelings of respect, status and progress "derived from how places are created, serviced and talked about by those who live there". The authors add that even a person's position in society in terms of "relative position and advancement, self-esteem and self-efficacy" (Ibid, 2012:11) are affected. The Chartered Institute of Housing (2013) additionally reports that housing is a key part of the physical, social and economic fabric.

Housing and the physical environment

The majority of new and existing housing in the United Kingdom purports significant environmental impacts (Environmental Audit Committee, 2005; WWF, 2003), including approximately 27% of carbon dioxide emissions in the United Kingdom, and has implications for climate change as well as significant impacts on timber and water consumption (Entec, 2004; WWF, 2003).

In terms of the environment, there are around 24 million homes in the United Kingdom today. It has been calculated that 80% of these will remain standing in 2050. New housing adds at most 1% to the existing housing stock, whilst the other 99% produce 27% of total UK carbon emissions, use half of all public water and generate 8% of total waste (Power, 2008). Because housing has such wide ranging impacts, it is important that it is taken into serious consideration as part of the overall subject area of urban regeneration.

Housing can improve the environment (Bhatti, 1999). A good example of this is the evidence on the positive impacts of housing programmes on both housing and neighbourhood quality, with improvement of the housing stock usually being accompanied by regeneration of the surrounding physical infrastructure and public spaces (Kleinhans, 2012). However, there have been persistent concerns raised over the quality of some of the housing in the United Kingdom (Maddedu, 2013). This could be related to the fact that the United Kingdom has one of the oldest housing stocks in Europe. Indeed, at current replacement rates, 85% of the housing stock that will be in use in the 2020s is already standing (Whitehead, 2004).

Various aspects of housing can have significant negative impacts on the environment, including "location, construction, design, maintenance, management, use and demolition" (Winston, 2010:319), whilst housing also has environmental importance as homes consume resources and generate waste and emissions (Carter, 2012). Additionally, construction consumes a considerable amount of other valuable environmental resources such as wood, minerals, energy and water (Huby, 1998).

Not only does the construction of housing use resources such as energy and water, but the use of housing consumes energy and water as well as the production of waste. In addition, the demolition of housing involves the production of high quantities of waste, some of which can potentially be toxic (Edwards and Turrent, 2000).

Housing and the social environment

Good housing can influence a person's quality of life, together with their life expectancy and the social opportunities available to them (Audit Commission, 2009). Indeed, it has been stated explicitly that "sustainable housing has a key role in the quality of human life" (UN, 2015). Housing can also convey a sense of pride, status, the foundation of family life, a hub for social networking with family and friends, the "frontline in the social support system" and the development of personal and cultural identity (Carter, 2012:68). Housing also provides a personal space for the individual, and in particular, "the place with which the occupant identifies a basic urban existence". It also provides a space for privacy and family life (Maliene and Malys, 2009).

There are many things that human beings would find difficult to do without good quality housing, such as find and keep a job, learn, maintain health, to vote, to claim benefits and to initiate and maintain stable relationships (King, 2012).

An additional advantage of housing, in a social context, is that through new construction, conversion of rental stock to owner-occupied housing and other strategies, the social mix of an area can be changed. Housing allocation strategies used to stimulate social interaction, based on common shared characteristics of future residents, can change an area's social make-up (Kleinhans, 2012:3). The same author also mentions the notion of empowerment, where individuals "gain control and influence over their lives and become democratically enabled to participate in society" (Ibid, 2012:4). Housing can provide this empowerment, notably through such examples as the sale of social or public housing to renters via the Right to Buy scheme.

In addition, the notion of repairing, altering and improving owner-occupied housing to the taste of the individual gives increased control which "contributes to a more general sense of control" over important life events (Rohe and Stegman, 1994), as well as providing a sense of security, as owner-occupiers cannot be evicted. The perceived improvement in status which comes about as a result of the purchase of a home is an additional social benefit to people, via the psychological theories of social comparison and self-attribution (Kleinhans, 2012). In terms of community development, it has been cited that this is at its most successful when housing-led regeneration occurs. Similarly, the emphasis of the sustainable community policy has always been on regeneration with housing at the core (Rowlands, 2010).

Housing and the economic context

Liu (2010) states that the United Kingdom's most valuable asset continues to be housing, with a total value of £4.3 billion, whilst the Office for National Statistics (2016) furthers this by claiming that the figure is even higher – at £5.5 trillion, this being equivalent to 62% of the nation's wealth (Ibid., 2016).

According to the House of Lords (2015), housing is very important for the economy of the United Kingdom. Indeed, the housing sector is an important economic sector which plays an important role in the promotion of economic growth and which additionally affects economic and social stability (Xue, 2012). The Audit Commission (2009) writes that local economies will only thrive if people who work in an area can find the right housing within easy reach of their jobs. Indeed, Clapham et al. (1996) have even suggested that through the encouragement of housing investments on different levels, it is possible to reduce unemployment both on a national and regional scale through the creation of new job opportunities.

The price of housing can also act as a mediator which links housing markets and the rest of the economy, as housing price fluctuations represent the main source of fluctuations in housing wealth (European Central Bank, 2003).

The inability to find housing has economic consequences. For example, those unable to find housing in areas of growth will be less able to participate in labour markets and to raise their human capital by doing so, whilst the construction and regeneration of housing additionally provides economic investment and can act as a driver for economic growth (Mah, 2012). Indeed, as Carter (2012) writes, investment in housing has traditionally been used to trigger job growth in times of high unemployment and during recessions in order to stimulate the economy.

Sustainable development

History of sustainable development

The terms sustainability and sustainable development are sometimes used interchangeably; however, Brandon and Lombardi (2011) opined that the term sustainable development is meaningful considering the root terms: *sustain* and *develop* which implies 'facilitating improvement without jeopardizing what already exists'. The most common cited definition is that of the report of the World Commission on Environment and Development, popularly known as the Brundtland Commission, as: 'the development that satisfies the needs of the present generation without compromising the ability of future generation to meet their needs' (WCED, 1987). It could be argued that the history of sustainable development goes back further than many commentators enlighten upon.

Some historical foundations of sustainable development have been traced back to Plato in the 5th century (Du Pisani, 2006), and even to the 18th, 19th and early 20th centuries (Holland, 2003; Lumley and Armstrong, 2004; Pepper, 1996), whilst others assert that it was specifically in 1798 that sustainable development was truly born, through the publishing of the works of Thomas Robert Malthus (1766–1834) and his *Essay on the Principle of Population*, which argued that there would come a time when food production would not be able to keep pace with the growth of population.

However, some authors write that it is possible to trace the roots of sustainable development back to the more recent time of the 1950s and 1960s, when developed nations were becoming increasingly aware that the environment was being affected by rapid industrialisation (Ashford and Hall, 2011; Hall and Ashford, 2012) or alternatively, only as far back as 1969 (Adams, 2006), or indeed the early and mid 1980s (Clark and Munn, 1986; IUCN/ UNEP/WWF/FAO/ UNESCO, 1980; World Commission on Environment and Development, 1987), or even specifically in 1987 (Mebratu, 1998). Du Pisani (2006) adds that the terms 'sustainability' and 'sustainable' appeared for the first time in the Oxford English Dictionary during the second half of the 20th century.

In the late 1960s, there were worries about environmental problems, rapid industrialisation and whether natural resources could meet the requirements of a growing population. Relevant publications citing these arguments include *The Tragedy of the Commons* (Hardin, 1968) and *The Population Bomb* (Ehrlich, 1968).

Subsequently, and along the same lines, in 1972, the Club of Rome and their book entitled *The Limits to Growth* described the effects and issues relating to non-renewable resource depletion and the urgent need for sustainability. In the same year, the UN Conference on the Human Environment was held in Stockholm, Sweden. This conference identified the need for all nation states to establish an environmental policy at national level and brought the importance of the environment for sustaining life to the world's attention (Caldwell and Weiland, 1996). The conference is credited as being a catalyst for the large scale of literature subsequently being published regarding the world's consciousness about the natural environment (Dernbach, 1998; Emmelin, 1972; Strong, 1972). Indeed, according to Atkinson (2008), the early debate on sustainable development mainly centred on environmental protection.

In 1983, the UN General Assembly created the World Commission on Environment and Development (WCED) (later known as the 'Brundtland Commission', named after its Chair, Gro Harlem Brundtland, the then Prime Minister of Norway), and in 1987, the WCED published the Brundtland Report, entitled *Our Common Future* (WCED, 1987) and within the report, seminally defined sustainable development as "development that meets the needs of the present without compromising the ability of future generations to meet their own needs" (WCED, 1987:43). It is this definition that is often referred to as the gospel definition of sustainable development.

The UN Conference on the Environment and Development (UNCED), held in Rio de Janeiro, Brazil, during the summer of 1992, then took the concept of sustainable development further by formulating the Agenda 21 'Policy plan for environment and sustainable development in the 21st century' which extended the ongoing debate beyond simply environmental issues, which had been the case up until that point in time (Ang and Wilkinson, 2008; Atkinson, 2008; Du Plessis, 2005) by reinforcing the integration of the 'three pillars of sustainable development'; namely economic development, social equity and environmental protection (Brandon and Lombardi, 2011; Carter and Rogers, 2008; Elkington et al., 2007).

Subsequently, in September 2000, at the Millennium Summit held in New York, world leaders agreed on the 'Millennium Development Goals'. These eight goals sought to demonstrate that "the livelihoods and well-being of the world's poor are now conceptualised in terms of access to opportunity and absence of insecurity and vulnerability" (Adger and Winkles, 2007:194).

In June 2012, the UN Conference on Sustainable Development (known as Rio+20) was held in Rio de Janeiro, Brazil, 20 years after the first Rio Conference. The primary purpose of the conference was to reinvigorate the promotion of sustainable development within the international community. It has widely been argued that the concept of sustainable development has evolved from a consideration of only environmental factors, through then to economic and also social considerations (Maliene et al., 2008). Hawkins and Shaw (2004) explain further that the ultimate goal of sustainable development is to ensure convergence among the three pillars of sustainability, or as it is alternatively and often termed, the triple bottom line (Elkington et al., 2007). The Rio+20 conference report, entitled *The Future We Want*, highlighted the 'green economy' as being one of the important tools available for achieving sustainable development (UN, 2012a:9). The 'green economy' was referred to within the UNEP Green Economy Report of 2011, as "one that results in improved human well-being and social equity, while significantly reducing environmental risks and ecological scarcities. It is low carbon, resource efficient, and socially inclusive" (UNEP, 2011).

In the context of the present day, the concept of sustainable development has been recently re-contextualised by the United Nations Conference on Housing

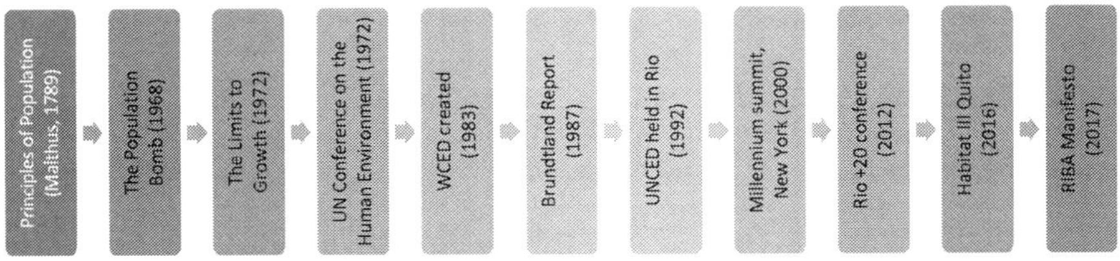

Figure 2.1 Timeline of publications related to sustainable development.

and Sustainable Urban Development (Habitat III), held in October 2016 in Quito, Ecuador, through the adoption of the 'New Urban Agenda', defined as

> a shared vision for a better and more sustainable future – one in which all people have equal rights and access to the benefits and opportunities that cities can offer, and in which the international community reconsiders the urban systems and physical form of our urban spaces to achieve this.
>
> (UN, 2017)

The concept of sustainably developing urban areas is related to social issues, including poverty, health, and housing. The General Assembly of the United Nations asserted in the resolution 'Transforming our World: the 2030 Agenda for Sustainable Development', which was adopted on 25 September 2015, that there would be 17 Sustainable Development Goals (SDGs), including, Goal 11 – "Make cities inclusive, safe, resilient and sustainable" (Dean et al., 2017a). This goal is closely linked to the main domain of this book, setting as its primary target the provision of sufficient affordable housing. In addition, in the context of the United Kingdom, the Royal Institute of British Architects' (RIBA) Manifesto of 2017 continues the focus on the concept of sustainable development but at the same time, changes course from the past by "introducing a requirement to consider social return" in the quest for sustainable development, thereby "following the same approach as the Social Value Act" (RIBA, 2017). It is from this stance that this book finds further development (Figure 2.1).

Definitions

There is little by way of a consistent definition of sustainable development within the literature. Indeed, the term has been said to have a complex range of ideas and meanings (Adams, 2006). Parkin (2000), for example, found more than 200 definitions of sustainable development, and according to Holmberg (1994), by 1994, there were more than 80 different definitions of the term. Roberts (2003) has cited that there has been both a historical misinterpretation of the term as being principally concerned with environmental issues with an additional tendency "for economic issues to dominate social and environmental concerns".

The most common definition of sustainable development, as already touched upon, is the one conceived at the World Commission on Environment and

Development (WCED, 1987): "Development that meets the needs of the present without compromising the ability of future generations to meet their own needs". Despite its acceptability, it is sometimes criticised for the lack of precision (Thomas and Adams, 1997). The definition centres on equity and futurity assumptions, but it is argued that the futurity and equity assumptions were not qualified. In addition to time, another of such imprecisions is the dimension of space. Some authors assert that the definition lacks distinction between needs and wants, as well as economic growth and development. Another deficiency is the absence of a mention of 'natural environment' despite its significance, whist giving emphasis to 'human needs' (Diesendorf, 2000). It does, however, highlight the triple bottom line (Elkington et al., 2007), together with the concept of needs and the importance of a longer term perspective, and the recognising of potential trade-offs between meeting current needs and ensuring that resources are available to meet future needs and implies a form of socio-economic development, increasing human wellbeing or quality of life for today's generations but not leading to a decline in future wellbeing (Ekins et al., 2008). Parkin (2000) echoes this capacity for continuance and asserts that "sustainable development is the process whereby over time, we achieve sustainability". The notion of the triple bottom line, or components of it, are involved in other definitions within the literature. Elkington et al. (2007) refer to sustainable development as containing all three aspects of the triple bottom line. This is sometimes referred to as the 'triangle' of 'People, Planet, and Profit' (the three Ps) (European Commission, 2002) or 'People, Planet and Prosperity' (Pope et al., 2004). Roberts and Benneworth (2002) reaffirm the notion of the triple bottom line in citing that "sustainable development balances economic, social and environmental development to maximise living standards for most people over time".

The social and economic systems are referenced within Pearce et al. (1989) and Pearce (1993), whilst economic and environmental aspects are referred to in HMSO (1994) and IUCN/UNEP/WWF (1991) refers to the social and the environmental.

In the UK Government's *A Better Quality of Life* report (DETR, 1999), the UK Government furthers the holistic stance, when asserting that sustainable development

> means meeting four objectives at the same time, in the UK and the world as a whole: social progress which recognises the needs of everyone, effective protection of the environment, prudent use of natural resources and maintenance of high and stable levels of economic growth and employment.

Models

Ekins et al. (2008) developed what they termed the 'four capital model of sustainable development'. This can be seen in Table 2.2.

Table 2.2 Four types of capital, taken from Ekins et al. (2008)

Type of capital	*Definition*
Manufactured capital	Manufactured (or human-made) capital is what is traditionally considered as capital: produced assets that are used to produce other goods and services. Some examples are machines, tools, buildings and infrastructure.
Natural capital	In addition to traditional natural resources, such as timber, water and energy and mineral reserves, natural capital includes natural assets that are not easily valued monetarily, such as biodiversity, endangered species and the ecosystems that perform ecological services (e.g. air and water filtration). Natural capital can be considered as the components of nature that can be linked directly or indirectly with human welfare.
Human capital	Human capital generally refers to the health, wellbeing and productive potential of individual people. Types of human capital include mental and physical health, education, motivation and work skills. These elements not only contribute to a happy, healthy society, but also improve the opportunities for economic development through a productive workforce.
Social capital	Social capital, like human capital, is related to human wellbeing, but on a societal rather than individual level. It consists of the social networks that support an efficient, cohesive society and facilitates social and intellectual interactions among its members. Social capital refers to those stocks of social trust, norms and networks that people can draw upon to solve common problems and create social cohesion. Examples of social capital include neighbourhood associations, civic organisations and co-operatives. The political and legal structures that promote political stability, democracy, government efficiency and social justice (all of which are good for productivity as well as being desirable in themselves) are also part of social capital.

Originally described by Paul Ekins in 1992, the 'four capital model' was developed by economists at The World Bank and then furthered by the UK sustainable development charity 'Forum for the Future', where the fifth capital, that of 'financial capital' was added.

The 'five capitals model' provides a basis for understanding sustainability in terms of the economic concept of wealth creation or 'capital' (Forum for the Future, 2017) and can be seen in Table 2.3.

There are additionally pictorial diagrams of the conceptual make up of sustainable development, including the typical 'Venn diagram' and the 'Russian Doll Model'; Figures 2.2–2.4.

According to the Venn diagram, the interaction between the three triple bottom line components of sustainable development is shown, with sustainable development being achieved where the three dimensions overlap in the diagram.

Table 2.3 The five capitals (taken from Parkin, 2000)

The five capitals
1 **Natural capital** (also referred to as environmental or ecological capital) represents the stock of environmentally provided assets and falls into two categories. a Resources, some of which are renewable (trees, vegetation, fish, water), some non-renewable (fossil fuels, minerals). In some places ostensibly renewable resources (such as fertile soil) have become non-renewable (desert). b Services, such as climate regulation or the powerful waste processing cycles.
2 **Human capital** consists of the health, knowledge, skills, motivation and spiritual ease of people. All the things that enable people to feel good about themselves, each other and to participate in society and contribute productively towards its wellbeing (wealth). Recently recognised as providing a high return on investment, especially in developing societies (where investment in human resources is viewed as possibly the most essential ingredient of development strategies) but also in the highly industrialised world.
3 **Social capital** is all the different cooperative systems and organisational frameworks people use to live and work together, such as families, communities, governments, businesses, schools, trade unions, voluntary groups. Although they involve different types of relationships and organisations, they are all structures or institutions that add value to human capital. Again, the importance of social capital has only recently been recognised, unfortunately however, there are increasingly visible negative effects when it is eroded.
4 **Manufactured capital** comprises all of the human fabricated 'infrastructure' that is already in existence: the tools, machines, roads, buildings in which we live and work, and so on. It does not include the goods and services that are produced, and in some cases, manufactured capital may be viewed as a source of materials (e.g. building waste used as aggregate for road building or repair).
5 **Financial capital** has, strictly speaking, no intrinsic value; whether in shares, bonds or banknotes, its value is purely representative of natural, human, social or manufactured capital. Financial capital is nevertheless very important, as it reflects the productive power of the other types of capital and enables them to be owned or traded.

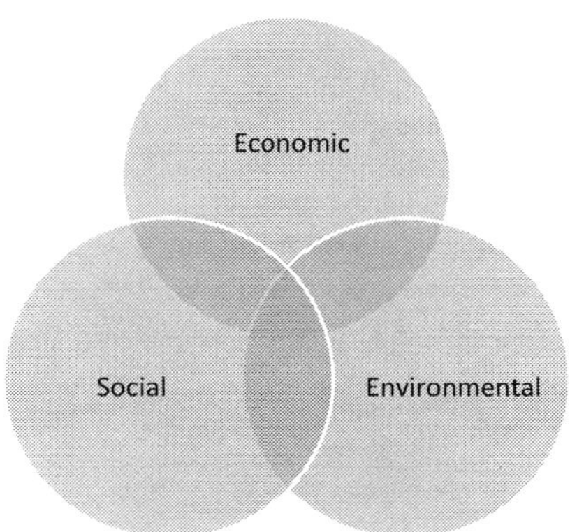

Figure 2.2 The Venn diagram of sustainable development (Adams, 2006).

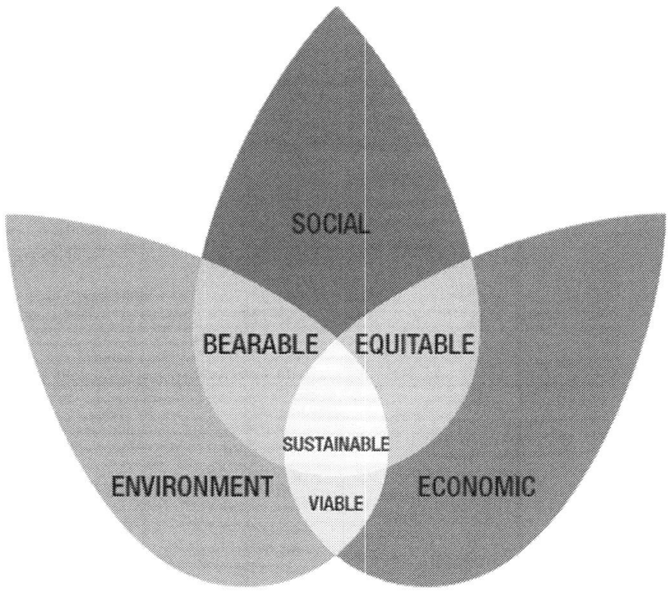

Figure 2.3 The Venn diagram of sustainable development showing the interaction of social, environmental and economic factors (Sullivan, 2012).

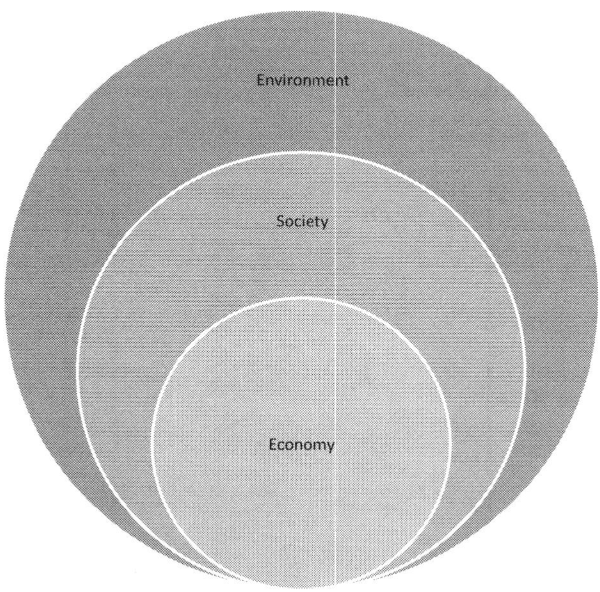

Figure 2.4 The 'Russian Doll Model' of sustainable development (O'Riordan et al., 2001).

The three pillars of sustainability may not necessarily be distinguishable from each other, as there is overlap. A sustainable development will seek to satisfy the socio-economic as well as the environmental requirements for sustainability. The interplay can be defined yet further, see Figure 2.3, adopted from Sullivan (2012).

Alternatively, a more altogether efficient way of presenting sustainable development is through the 'Russian Doll Model' (O'Riordan et al., 2001). This model depicts the fact that economic activity should lean towards social progress and should be achieved within environmental limits:

Perspectives

Improved standard of living for sustainable development requires economic growth, however, the call for a 5–10 fold increase in economic growth in the Brundtland Report (WCED, 1987) is perceived to be in support of a toxic economic growth in the use of materials and energy at the expense of the natural environment (Diesendorf, 2000). When the WCED reconvened in 1992, this call was abandoned despite being the basis for the outcome of the Rio Earth Summit held in 1992 in Brazil (Brandon and Lombardi, 2011). This suggests that in sustainability thinking, however much economic growth may be necessary, it must not compromise the integrity of the physical environment.

'Strong' and 'weak' sustainability and different evaluative perspectives

There is reference within the literature to two models of sustainability; those of 'strong' sustainability and of 'weak' sustainability.

Strong sustainability takes the view that there is imperfect knowledge about the overall contribution of environmental capital, and that the environment has not been correctly assigned value (Ehrlich and Ehrlich, 1992), with the consequence that environmental capital should be protected, as some environmental capital may be 'non-substitutable' (Turner, 1993). A strong sustainability position holds that fundamental services provided by nature and elements of natural capital are critical because of their unique contribution to human wellbeing (Ekins et al., 2003). Most ecologists and environmentalists are proponents of the strong sustainability stance (Stoddart, 2011).

However, a weak sustainability stance assumes that natural and manufactured capital are both substitutable and that there are essentially no differences between the kinds of wellbeing they generate (Ekins et al., 2003; Neumayer, 2003, 2012). As Solow (1993) explains, within this model of understanding, the only thing that would matter is the total value of the aggregated stock of capital (Ibid, 1993) and the view would be held that the total stock of capital assets would be perfectly substitutable for one another. As Neumayer (2003:1) explains, "it [therefore] does not matter whether the current generation uses up nonrenewable resources or dumps Carbon Dioxide in the atmosphere, as long as enough machineries, roads and ports are built in compensation". According to this view, substitution can take place between natural and man-made capital (Pearce and Atkinson, 1992).

The basic distinction between 'weak' and 'strong' sustainability has important implications as to whether environmental systems and resources should be kept intact by themselves or, whether contrary to this, the environment would be allowed to decline as long as the overall value of society's economic capital is kept intact and of course they also have important implications on the conceptual framework underpinned in the socio-environmental balance of costs and mitigation measures.

Further to this distinction, and in relation to the implications on evaluative models derived from different views on sustainability, it may be worth mentioning how the current coexistence of multiple paradigms on ecology has an influence on the philosophical stance underpinned in the evaluative process. According to the supporters of a vision of ecology referred to as *deep ecology* (Naess, 1973), the attribution of a monetary value to something which should have an intrinsic existence value would clash with the rationale to preserve it – regardless of its cost – for future generations, in line with the principle of intergenerational justice. With respect to this position, this research assumes that the act of monetising the value of environmental and social costs and benefits is not an ontological declaration and does not imply assimilating non-market goods to market goods. This is merely instrumental to enable the raising of consensus (and financial support) across as many stakeholders as possible regarding the need for wider interventions relating to a crucial pillar of the SDGs, which is, the allowing of each person the dignity of a decent home, by unveiling the larger and long-term impacts on respective budgets.

Sustainable development and its links with urban regeneration

With sustainable development incorporating the notion of the triple bottom line (Brandon and Lombardi, 2011; Carter and Fortune, 2007; Carter and Rogers, 2008; Drakakis-Smith, 1995; Hawkins and Shaw, 2004; Rodriguez et al., 2002), it could be argued that the two concepts of urban regeneration and sustainable development share the same fundamental characteristics and can thereby be linked as a consequence. This link is highlighted within the literature. Jones and Evans (2008) refer to the concept of sustainability as being "central" to regeneration policy in the United Kingdom. Couch and Dennemann (2000) state that the strategic aims of urban regeneration are "amenable" to those of sustainable development in a key way and additionally argue that all urban regeneration contributes to sustainable development, through the recycling of derelict land and buildings, reducing demand for peripheral development and facilitating the development of more compact cities, whilst Boyko et al. (2012) write about the "key role" urban regeneration plays in sustainable development. Zheng et al. (2014:1) add that "as sustainable development corresponds to urban renewal in terms of social, economic and environmental sustainability, it has been recognised that urban renewal and sustainability should be combined together". Pitt et al. (2009)

assert that anyone associated with creating the built environment has a key role to play in delivering and ensuring the sustainability of communities.

The DETR (the UK's Department of the Environment, Transport and the Regions; DETR, 2000) and ODPM (the UK's Office of the Deputy Prime Minister, this being replaced by a new Department for Communities and Local Government in 2020) add that sustainability is at the core of UK regeneration strategies (ODPM, 2003, 2005); the United Kingdom being one of the first developed nations to produce a national strategy on the sustainable development and regeneration concept, taking the lead to implement many sustainable development and regeneration principles and many of the Agenda 21 action plan (the action plan was derived from the United Nations Conference on Environment and Development in 1992), in theoretical and practical terms (DEFRA, 2011).

It has been stated that a poorly planned and designed sustainable regeneration project can form the basis of social conflict (Yau and Chan, 2008), whilst many policy systems and strategies have been developed over the years in order that sustainable development and regeneration are realised (Lam et al., 2011; Van Bueren and De Jong, 2007). The notion of sustainability has been actively and widely promoted within sustainable regeneration projects, but apparently with limited sustainability benefits for the intended beneficiaries (Van Bueren and De Jong, 2007).

In addition, the built environment is known to represent the physical structure and the vehicle for sustainable development (Brandon and Lombardi, 2011), whilst built environment professionals have been urged to recognise the importance of sustainable development as an integral aspect of built environment practices (Bennett and Crudgington, 2003). It is also argued that any regeneration project that fails to evaluate each of the established sustainability pillars will not achieve its sustainable development and regeneration objectives (CLG, 2008; Winston, 2009).

Summary

This chapter has provided an introduction to the subject domains of sustainable development and urban regeneration and has covered the history and definitions of sustainable development, together with definitions of 'weak' versus 'strong' sustainability, models of sustainable development, 'capitals', the UN Habitat III conference and its take on sustainable development and sustainable development and its links with urban regeneration. Further to this, the history and definitions of urban regeneration were covered. The chapter then proceeded to focus on the importance of housing within urban regeneration, including the many links between housing and urban regeneration, housing and health, as well as housing and its relationship with each aspect of the triple bottom line (Elkington et al., 2007).

Access to housing is considered as a human right (Universal Declaration of Human Rights, 1948) which has a multidimensional impact on human life (UN, 2016). In addition, because housing has been referred to as possessing "social, environmental and economic dimensions which are all closely interrelated", it is cited as being "much more than providing people with a place to live" (UN, 2016).

Housing can additionally be said to play a 'pivotal role' in the regeneration processes being carried out in urban areas (Ha, 2007:118) and can also be considered to be a driving force of many regeneration schemes (Haran et al., 2011) or as being vitally important (ODPM, 2003). It has been referred to as being one of the most important public policies affecting urban development and consequently has significant potential to contribute to sustainable development (Priemus and ten Heuvelhof, 2005; Tosics, 2004).

It is within the housing-led regeneration area that there is a particular niche for a solution to address the lack of a scientific, quantitative and accurate evaluation method of measuring socio-environmental impacts. Indeed, some housing-led organisations do not even take such impacts on board. With the economic cuts faced by such organisations, it is increasingly important that the above-mentioned impacts are measured, and measured correctly.

The economic climate and its effect on housing associations covered in Chapter 1 is also re-casted as a serious challenge to the achievement of SDGs, since failures in ensuring sufficient support to the housing sector would result in hindering a main pillar of the New Urban Agenda. The findings and solutions provided within this book can therefore be taken as a valuable contribution to the achievement of the international goals set by Habitat III.

3 Evaluation

Its background and history within urban regeneration

Introduction

This chapter will cover the subject of evaluation, its definitions and history. A brief history of urban regeneration evaluation in the UK context will be outlined, together with a background to evaluation within urban regeneration more generally. Finally, a critical review of current assessment methodology including that of the EGRUP Guidance, City Challenge, the Single Regeneration Budget, the New Deal for Communities, Urban Development Corporations, the Hemphill Framework, the Sustainable Urban Renewal Project Assessment Model (SURPAM), the SURegen Workbench, LUDA, the UK Government 'Green Book' and the Royal Institution of Chartered Surveyors (RICS) 'Red Book' will be carried out.

Evaluation

Evaluation can be defined as the "systematic examination and assessment of the features of an initiative and its effects in order to produce information that can be used by those who have an interest in its improvement or effectiveness" (WHO, 1998) and in its broadest form, "means to ascertain the worth of or to fix a value on some object" (Rossi et al., 2004). The 'value on offer' has particular prominence in the current economic climate, ensuring that decision making and appraisal is part of economic accountability (Raphael, 2000).

In a more holistic sense, evaluation is defined by Smith (2006) as "the systematic exploration and judgement of working processes, experiences and outcomes… [which] pays special attention to aims, values, perceptions, needs and resources".

A brief history of evaluation

Scriven (1996) states that although evaluation is still young in terms of being a subject discipline, in practical terms however, it can be classified as being quite historic. In this regard, Guba and Lincoln (1981) cite personnel selection tests being carried out in China in 2200 BC by the incumbent Emperor as one of the earliest known examples of evaluation taking place. More recently, it has been declared that the historical roots of evaluation can be found as far back as

DOI: 10.1201/9781003043881-3

the 17th century (Rossi et al., 2004), or, alternatively and specifically in the year 1792 when William Farish invented the quantitative mark to score examinations (Hoskins, 1968).

Madaus and Kellaghan (1982, cited in Hogan, 2007) state that in a similar way at that time, education in the United Kingdom was also reformed through evaluation. Subsequently, Stufflebeam and Shinkfield (2007) go on to state that the first systematic school surveys "using printed tests" were conceived and produced in Boston, USA, and were used to test students area-wide, across a range of subject areas.

The year 1815 brought about the earliest 'formal' evaluation in the form of the US Army's standardised production policies for materials and products (Smith, 1987, cited in Hogan, 2007).

In the late 19th century, the educational reformer of New York, Joseph Mayer Rice (1857–1934) conducted what is generally recognised to be the first formal education programme evaluation in the USA (Stufflebeam and Shinkfield, 2007) when attempting to analyse students' spelling deficiencies (Hogan, 2007).

The 20th century brought about differing approaches to evaluation, including one particular school of thought introduced by Frederick Taylor, a leading US manufacturer. The standardisation and subsequent efficiency and consistency of quality used in that field influenced education leaders at the time to seek the same level of standardisation and efficiency in schools (Ireh, 2016). The standardised testing system was born and was widely used throughout the United States and the United Kingdom.

There are examples of systematic evaluation being carried out as part of First World War assessment exercises carried out by the US Government (Rossi et al., 2004), whilst early evaluators were described by Freeman (1977) as social scientists who used "rigorous research methods to assess social programmes in a variety of areas".

In the 1930s and 1940s, Dr Ralph Tyler, one of the leading educational researchers of the 20th century and a man who was an educational advisor to six US Presidents (Stanford University, 1994) brought about an approach which was based on clearly stated objectives. During the Second World War, pioneering evaluations of the attitudes of American Soldiers were carried out (Stouffer et al., 1949), with similar research being carried out in Britain and elsewhere around the world (Freeman, 1977). The 1950s continued in much the same way, with programme evaluation being carried out as a part of programmes within "urban development and housing, technological and cultural education, occupational training and preventative health activities" (Rossi et al., 2004).

The 1960s brought about the birth of criterion reference testing (Hogan, 2007), whilst Scriven, Stufflebeam and Stake introduced new models in the 1960s and 1970s which "departed radically from prior approaches" (Stufflebeam and Shinkfield, 2007). The 1960s also saw an increased number of books and papers published as evaluation research grew, whilst the 1970s saw the launch of the first journal dedicated to evaluation research; that of *Evaluation Review*, established in 1976 (Rossi et al., 2004) and the popularity of the subject continued into the

1980s and beyond. In later years, standards are the latest measures to have been brought into the evaluation area, including those within the International Handbook of Educational Evaluation (Kellaghan and Stufflebeam, 2003).

According to Madaus et al. (2000, cited in Hogan, 2007), there are seven phases within the historical context of evaluation:

- Age of reform (prior to 1900)
- Age of efficiency (1900–1930)
- Tylerian age (1930–1945)
- Age of innocence (1946–1957)
- Age of development (1958–1972)
- Age of professionalisation (1973–1983)
- Age of expansion and integration (1983–2000)

Even in terms of the mere reporting of the history of evaluation, differences in opinion are highlighted. Stufflebeam and Shinkfield (2007) only identify with five periods:

- Pre-Tylerian period (prior to 1930)
- Tylerian age (1930–1945)
- Age of innocence (1946–1957)
- Age of realism (1958–1972)
- Age of professionalisation (1973 to present day)

A brief history of urban regeneration evaluation in the United Kingdom

Since the mid-1980s, the British Government has "put more emphasis on evaluation and monitoring" (Ho, 1999). The UK Government concurs, stating that "regeneration has had a reasonably strong tradition of appraisal" (CLG, 2008).

This is however contradicted by both the CEA (2001), which states that the only UK guidance document specifically focused on evaluation in urban regeneration is known as the *EGRUP* (Evaluation of Regeneration Projects and Programmes) guidance document (CEA, 2001) and additionally by Pollitt (1993), which states that evaluation in the United Kingdom has tended to be a poor relation. It reads that "the history of the last thirty years reveals that policy evaluation has never found a secure or permanent home near the heart of a [relatively centralised] state machine".

The UK Parliament (2003) expressed concerns regarding the lack of intellectual sophistication regarding evaluation methodology. It appears that despite many previous British evaluation studies being carried out, there still exists an inadequate understanding of how and why programmes are or were successful or not (Ho, 1999). A main assumption in the past has been that by simply pumping money into an area, this would solve any inherent problems (Ho, 1999). Though this policy approach was stemming from the Keynesian theoretical model of the economic multiplier applied to public policies, a variety of issues undermine this assumption. This has been also contradicted by the UK Department of the Environment, Transport and the Regions (DETR). The DETR expressed that deprivation in funded areas has not improved despite continued investment in pinpointed areas (DETR, 1998). Indeed, Robson et al. (1994) state that between 1979 and 1991, a total of some £900 billion was spent on urban regeneration initiatives in the United Kingdom and that to this day, money is still being channelled into the very same areas.

Continuing in this school of thought, Imrie and Thomas (1995) write that the intensity in which new urban policy initiatives have been created since the early 1980s has not been matched by any form of systematic monitoring and evaluation which could advise on where Government or other organisations have been potentially going right or wrong in their approaches, whilst Turok (1989) adds that evaluative research has tended to be illogical, weak and increasingly concerned with a narrow range of quantitative evaluation criteria.

Commenting chronologically, the United Kingdom has some history of strategic evaluation as early as the 1970s through such organisations as the Central Policy Review Staff (CPRS) (Pollitt, 1993). However, the CPRS was disbanded due to the economic problems of that decade and because of the fall from power of the Conservative Government at the time. In addition to this, the 1970s witnessed a major restructuring of local government, which had a great impact in terms of corporate planning (HMSO, 1972).

The 1980s saw the creation of the National Audit Office and Audit Commission under Baroness Thatcher's Government and there was a criticism of neglect in terms of evaluation. This resulted primarily in the document *Policy Evaluation: A Guide for Managers* being produced and subsequently published (HM Treasury, 1988). Additionally, the UK Department of the Environment (DoE) produced *Improving Urban Areas: Good Practices in Urban Regeneration* (1988).

Expenditure cutting continued to be necessary in the 1980s and the stance taken by the then Conservative Government on this issue lead to a 'value for money' ideology being adopted within evaluation schemes. This lead to the birth of the so-called 'Output and Performance Measures' (Lewis and Jones, 1990). Indeed, according to Pollitt (1993), there was perhaps too much focus on value for money at the expense of any real evaluation taking place.

What also seemed to be lacking was quality or organisation in terms of Governmental approach, which then filtered down to local approach. For example, the *Urban Programme* of the 1980s was a major regeneration initiative of the time. This was replaced by *City Challenge* two years before its official evaluation assessment was carried out (Robson et al., 1994, cited in Ho, 1999). Subsequently to this, City Challenge was then replaced by the *Single Regeneration Budget*, two years before the interim assessment on City Challenge was completed (Russell et al., 1996). *New Deal for Communities* was introduced in 1998 to run alongside the Challenge Fund, when the evaluation study was only half completed (Brennan et al., 1998). Shaw (1995) adds that there was a marked absence of detailed government evaluations and policy reviews in the 1980s.

The 1990s saw evaluations that were to offer more "value" (Pollitt, 1993). Examples are cited such as Quality Adjusted Life Year (a measurement of disease burden), being used to allocate health resources and the Polytechnics and Colleges Funding Council actively using value for money evaluations. According to Ho (1999), "all new initiatives in the 1990s were subjected to national evaluation studies", however an engendering theme is that according to the author, "only limited lessons can be drawn from these studies". Ho (1998) states that "accumulation of knowledge was limited".

The 1990s saw the election of the New Labour Government, which continued with the same value for money theme. In wanting to be distanced from the previous Government, however, this was then re-termed "best value" (Department of the Environment, Transport and the Regions, 1998). This continued into the 1990s. Government documentation in 1998 regarding the New Deal for Communities Scheme advised that focus should be on whether value for money is being achieved with the public purse (Department of the Environment, Transport and the Regions, 1998). Ho (1999) seconds this, adding that "all regeneration initiatives in the 1980s and 1990s included stringent measures to record expenditure and output".

In the 1980s and 1990s, Ho (1999) adds that baseline data and quantitative information gathering was prevalent. However Julnes et al. (1998) repeat an earlier negative, by stating that the monitoring processes of initiatives at the time offered very little.

The post-1997 period was characterised by 'realist' approaches to evaluation that sought both to describe and explain policy outcomes (Ho, 1999; Taylor and Balloch, 2005). In 1998, the *New Deal for Communities* (NDC) was brought in which brought about the notion of partnerships working to regenerate areas. The official NDC report written by the Government (Lawless et al., 2010) recommended that the setting of realistic targets, prioritising places for regeneration, partnership working, time frames, spatial remits, utilisation of a "year zero" to properly set up a scheme, engaging with communities and sustainability were among the effective recommendations provided.

Subsequently, in 1999, it was the DETR which took the lead and produced the *Local Evaluation for Regeneration Partnerships: Good Practice Guide*. The guide provides a seven-stage process for evaluating regeneration as follows (UK Government, 1999):

1 Establishing and choosing effective base line indicators;
2 Obtaining and updating baseline information;
3 Establishing targets for comparative information analysis;
4 Adding value to output monitoring;
5 Monitoring changes in local conditions;
6 Assessing impact; and
7 The scheme evaluation report.

On a general note, evaluation methodology, according to Evans and Shaw (2001a, 2001b), did develop substantially in the 1990s, drawing particularly from environmental health (Bowling, 1997), crime prevention, urban design and quality of life measurement (Rapley, 2001).

National evaluations of the New Deal for Communities programme (CLG, 2008), Neighbourhood Management (CLG, 2008) and Local Strategic Partnerships (ODPM, 2005a, 2005b) in the 2000s showed that long-term national evaluations were starting to become more and more common place. It would appear that more research on areas including data on residents was indeed being carried out and was being used to formulate policy. However, Ho (1999) states that "the current British approach to evaluation is [still] based on the ideology of value for money".

In 2004, and coming towards the present day, the UK Government published documentation on the so-called '3R' interventions entitled *Assessing the impacts of spatial interventions: Regeneration, renewal and regional difference – 'The 3Rs Guidance'* covering regeneration interventions, renewal interventions and regional development policies whereby the ODPM (2004) sets out a broad framework for the evaluation of UK Government–backed regeneration projects. This framework utilises the notion of indicators as a measurement tool. There have also been Government-led evaluations carried out towards the present day, on such urban regeneration initiatives as City Challenge, the Single Regeneration Budget and the New Deal for Communities, which will be looked at in more depth later in this chapter.

Background to evaluation within urban regeneration

Although the evaluation of regeneration schemes has been referred to as a "vital task" (Roberts and Sykes, 2000), there does not appear to be an agreed consensus on how this should be carried out. Indeed, Bowey (1997) adds that although there is a need for monitoring and evaluation of regeneration initiatives, there has been "scant regard" paid to this aspect.

In addition, it would seem that any evaluations carried out have not had any real effect on wider progress within the field of urban regeneration. Carley and Kirk (1998) state that the need for urban regeneration hasn't been reduced, even after 30 years of policy initiatives.

Various authors and scholars have, over the years, published their varying methods, models, frameworks and metrics on how they feel measurement and evaluation should be carried out. Authors have highlighted important aspects of an urban regeneration development; for example, Jeffrey and Pounder (2000) highlight the importance of physical or aesthetic redevelopment; economic redevelopment is highlighted by such authors as McGregor and McConnachie (1995), and Colantonio and Dixon (2010) refer to social impact, but what other aspects exist as part of a successful, or unsuccessful programme? As Tyler et al. (2013) state, "it is an obvious question to ask why there is so little evidence available on the aggregate value of regeneration benefits in the light of the resources that many countries commit to them".

Further to this notion of evidence, Brandon and Lombardi (2011) assert that sustainability appraisal frameworks used in the built environment must address economic, environmental and social impacts to ensure continued sustainable development after regeneration. However, there have been found to be gaps in sustainability frameworks in practice (Clapham, 2014), and it has been argued that many frameworks sacrifice social and economic factors at the expense of the environmental (Carter and Fortune, 2007) with some assessment methods being described as either too diverse from other methods (Russell et al., 2000), inadequate (Smith, 2006), lacking (Madlener et al., 2003) or even concerning (Diamond et al., 2006).

Within the housing sector, the Homes and Communities Agency has demanded "robust assessments" of assets (HCA, 2012), and in addition, it has been cited that the schemes' value to society needs to be clear in today's challenging economic environment (Tyler et al., 2013). Although the need to take into account both quantitative and more qualitative or intangible impacts (Church and Young, 2001; Lee and Chan, 2008) has been cited (CLG, 2009; McQuaid et al., 2006), impact measurement has remained "under theorised" (Ebrahim and Rangan, 2010).

To this end, whilst there is as yet no universally wide accepted methodology of evaluation in urban regeneration (Bell and Morse, 2003), an alternative of traditional forms of evaluation would be to quantify or monetise the value of impacts normally categorised as being intangible; something previously considered difficult using conventional techniques (Vardakoulias, 2013). Such an alternative has, however, thus far had very little critique within the existing literature (Bichard, 2015; Watson and Whitley, 2016).

Social Return on Investment (SROI) has been used as the primary mechanism for evidencing organisational impact in this way, being endorsed by both the Cabinet Office and the Scottish Government (Watson and Whitley, 2016), whilst Bridgman et al. (2016) and Watson and Whitley (2016) have remarked on the potential usefulness of the SROI methodology within the field of the built environment. SROI has been taken further by Bichard (2015), creating Sustainable Return on Investment (SuROI) after being refined, proving that monetary evaluation can indeed be implemented to evaluate sustainable performance of projects within the built environment.

Previous methods

Critical review of assessment methodologies

This section will set the context for understanding how the assessment of urban regeneration strategies has been typically performed so far; it includes recent and/or current assessment methods commonly used to assess the impacts of investment in housing in different programmes or initiatives. The following methods, mainly related to governmental schemes, will be reviewed: (1) EGRUP Guidance,

(2) City Challenge, (3) Single Regeneration Budget, (4) New Deal for Communities, (5) Urban Development Corporations, (6) Enterprise Zones, complemented by the following methods: (7) Hemphill Framework, (8) Sustainable Urban Renewal Project Assessment Model, (9) SURegen Workbench, (10) LUDA, (11) UK Government 'Green Book' and (12) Royal Institution of Chartered Surveyors (RICS) 'Red Book'.

EGRUP guidance

The EGRUP (Evaluation of Regeneration Projects and Programmes) guidance uses a framework for the ex-post evaluation of expenditure and regeneration schemes (HM Treasury, 1995) and improves comparability of information on the value for money of regeneration programmes (Ibid, 1995), reinforcing the emphasis of Government at the time on value for money and the public purse (Gregory and Martin, 1988). It measures cost effectiveness via a basic cost benefit account for each regeneration measure concerned. EGRUP uses four proformas within its guidance, involving general management information about a scheme, quantifiable outputs and their costs, other relevant outputs and their costs and a measurement of additionality. It focuses on economic efficiency and outlines the inputs, outputs and outcomes of an urban regeneration scheme. The guidance refers to the importance of using indicators relating to government expenditure but concedes in section three of the guidance, that although all costs can be quantified, some "important" or intangible benefits cannot be. Such benefits, for example, social benefits, are simply listed, rather than being assigned a monetary and quantifiable value. This translates to such impacts not being recorded. In addition, attention is drawn to users of the guidance that there are issues with "outputs that are essentially unquantifiable" (HM Treasury, 1995). Environmental benefits were measured quantitatively, but by using a Likert scale (a psychometric scale whereby respondents indicate their agreement or disagreement to a statement). The guidance itself states that there is "clearly a need for a more rigorous approach, which would involve assigning monetary values to the environmental improvements" (Ibid, 1995).

It is consistently stated within the evaluation document that further research is required in the area of valuation for regeneration activities for which market-based information is not available.

City Challenge

The City Challenge (Russell et al., 1996) was commissioned by the DETR (UK's Department of the Environment, Transport and the Regions) and aimed to regenerate 31 areas through regeneration partnerships which ran in deprived urban areas between 1991 and 1998, with the aim of improving specific rundown inner city areas and improving the quality of life of local residents (DETR, 1998). In terms of the evaluation of this scheme, negatives include that there is no one method or framework in place to cover the multitude of partnerships which makes consistency of evaluation very difficult. The Final Evaluation of City Challenge (Russell et al., 2000) found that there was an enormous diversity of local evaluations which translated to an insufficient level of consistency to make meaningful comparisons across all areas. A more standardised methodological approach was considered to have been of wider value (DETR, 1998). Additionally, the evaluations undertaken or commissioned by the partnerships themselves varied significantly in quality, content and approach, with some choosing to assess the effectiveness, efficiency and impact of a programme, while others concentrated on updating baseline indicators (DETR, 1998). A key aspect of the schemes supported through the City Challenge Programme was their integrated nature – with traditional housing improvements being complemented by other regeneration projects concerned, for example, with job creation, training and crime and community safety, however, no method was set up to capture any of this information in numerical and quantifiable format. The DETR (1998) touches on the notion of sustainable development, including a statement that if a regeneration programme is to be regarded as successful, the outcomes it generates must be sustainable in some form, however, no method of measuring such aspects of sustainability is listed. Along the same lines, concern was also expressed within the City Challenge Final Evaluation on the lack of monitoring requirements for the quality of outputs or outcomes. Both City Challenge reports focus on outputs (rather than outcomes or impact) and focus on expenditure and outputs in relation to annual targets (OECD, 2004). Impact indicators form an integral part of the evaluation approach, but the indicators were chosen by the DoE, rather than stakeholders involved in the scheme in question (Ho, 2003). Further to this, it is cited by Robson et al. (1994) that some indicators, although appearing to be covering outcomes, are actually again measurements of outputs. The focus of the Government at the time was rather a case of what had been done, rather than what impact was had on local communities, meaning that City Challenge was in fact output oriented (Ho, 2003). This emphasis is shown in the listings of such items as the amount of jobs created, the amount of business and commercial floorspace built or the amount of dwellings improved (DETR, 1998).

The indicators used within the evaluation can be seen in Table 3.1. It is noticeable that the indicators are only available for specifically quantified aspects. Within the qualitative and more intangible areas, such as the *quality of life* category, it is stated that there is no standard indicator. In addition, within the unit cost per output statistics (DETR, 2000:179), values pertaining to *health* and *community and social* are simply listed without any figures as "not applicable" and impacts are simply not measured within the summary cost benefit account (Ibid, 2000). It is additionally admitted within the official final evaluation that "more emphasis should be given to assessing outcomes" (Ibid, 2000). It is this difficulty of quantification in terms of intangibles where the evaluation of City Challenge is unable to pick this area of impact up and where a methodology such as the previously cited SuROI can fill a gap in the knowledge base.

Table 3.1 City Challenge core impact indicators (*Dept of the Environment (1993) City Challenge Guidance Note: Appraisal, Monitoring and Evaluation, document issued to City Challenge Partnerships, London: DoE. Annex B.*)

City Challenge core impact indicators
a Employment/income Number of unemployed Number of residents of the City Challenge area obtaining permanent employment as a consequence of City Challenge actions Heads of households in receipt of income support
b Education Number of pupils entered for GCSEs achieving grades A to C in 5 or more subjects and number of pupils leaving school with no GCSEs
c Housing Number of households in owner occupation Number of unfit dwellings
d Environment/economy Hectares of derelict land and vacant land Floorspace in derelict buildings and vacant buildings
e Crime Reported crime per 1,000 households in residential burglary and theft of or from cars
f Quality of life/perception of City Challenge area No standard indicator but people's attitudes towards a range of issues should be tracked over five years

Single Regeneration Budget

The Single Regeneration Budget (SRB) (Brennan et al., 1997) used a cost benefit, inputs versus outputs approach within the study. Outcomes were measured in terms of indicators in areas such as jobs created or safeguarded, enhanced pupil attainment and personal development of young people and community safety initiatives (Zuberi and Taylor, 2013). There were three levels of outcomes involved within the evaluation. These levels cut across the distinction between social, economic and environmental goals and were termed delivery (the level of outputs), impact (the level of outcomes) and sustainability, termed as the longer term social, economic and environmental impact over the life of a given partnership. Indicators were used to gauge economic benefits, housing benefits, social benefits, environmental benefits and community benefits (Brennan et al., 1999), whilst an extensive baseline was also used (Rhodes et al., 2007). Social surveys and structured interviews were used to gauge opinions of key stakeholders within the SRB partnerships, together with an in-depth case study approach (Rhodes et al., 2007). The methodology looks to cite overall net additional achievements, however it is not able to numerically quantify the holistic sustainable impact of the programme in terms of all three aspects of the triple bottom line. In addition, it is stated that there are problems of comparability in the way the SRB is evaluated (Department of the Environment, 1997).

Perceptions about physical and environmental quality and amenity were assessed through resident and visitor surveys according to the methodology of Glennerster and Turner (1993), whereby scores were derived from direct observation (Department of the Environment, 1997). In addition, in terms of the measurement of social aspects, such as quality of life, quantified outputs in this area simply enumerate the numbers of new health, sports and cultural facilities and the numbers of local people given access to them. It is also stated that it is not thought generally possible to quantify all social effects (Ibid, 1997).

New Deal for Communities (NDC)

Each New Deal for Communities (NDC) programme was expected to achieve positive change in relation to six outcomes. Three of the outcomes were intended to improve 39 places. These were crime, the local community and housing and the physical environment. Three were to improve outcomes for people: education, health and worklessness (Lawless, 2011). A number of data collation and analysis tasks were central to the national evaluation, the most important of which was the biennial household survey. In 2002, a baseline was established across all 39 NDC areas using a survey questionnaire. This addressed socio-demographic, status and attitudinal considerations across all outcome areas. It was based on a random-sample survey design and used 3, 4 or 5 point Likert scales to quantify the responses (Lawless et al., 2010). Any change in NDC areas was benchmarked against other deprived, comparator areas. No previous evaluation of any English Area Based Initiative at that time had been able to explore questions of net change across all relevant regeneration areas and their residents, for all outcomes, from a common baseline (Lawless, 2011). NDC used shadow pricing methods to determine value for money, the first time that this had ever been done (Lawless et al., 2010).

Urban Development Corporations (UDCs)

According to Imrie and Thomas (1999), UDC evaluations are performance-related with a limited range of criteria, such as "jobs created and safe-guarded, hectares reclaimed and quantities of constructed roadway". The emphasis in UDC evaluations is on value for money with many commentators asserting that a greater range of non-quantifiable variables should have been accounted for within the scope of the evaluation (Hambleton and Thomas, 1995). Oatley (1989) states that performance indicators [were] mainly input and output measures and "did not provide a complete basis for assessment of how effectively the corporations have achieved their regeneration objective". A list of such indicators can be seen on page 10 of Oatley (1989) showing that social and environmental outcomes are not taken into account via a quantified methodology.

Enterprise Zones (EZs)

The final evaluation of the original EZs was carried out in 1995 (HMSO, 1995) and assessed the extent to which EZs had generated both additional economic activity and physical regeneration (Potter and Moore, 2000). The evaluation methodology covered an analysis of data, which included annual monitoring data provided by the Department of the Environment, in combination with further data collected by consultants to quantify the additional economic activity generated. Main factors analysed included employment characteristics, the number of firms established, industrial compositions of the firms within the zones, environmental improvements and impact on the local property markets (HMSO, 1995). The amount of jobs and costs per job created were highlighted, but the evaluation made a limited assessment of any inward investment to the zones (Potter and Moore, 2000). Face-to-face and postal surveys were sent out to local companies to gauge company perceptions of EZ benefits. Lastly, studies were carried out to assess the effects of EZ policy on local property markets, the creation of new economic activity and impact on the physical environment. There was simply no mention at all within archived governmental documentation on Enterprise Zones of social and environmental impacts being quantified.

Hemphill Framework

The Hemphill Framework (Hemphill et al., 2004) measures sustainability by allocating a points score to indicators within five areas, including: economy and work, resource use, buildings and land use, transport and mobility and community benefits. Residents and other users within a regeneration area are consulted through questionnaires and structured interviews where results contribute directly to a points scoring system. Points are allocated to each involved indicator and to each indicator set. Indicators are used to measure the performance of the regeneration scheme against sustainability criteria. However, the authors add that

> although it is possible to set indicator parameters for certain regeneration outputs (number of jobs created; amount of private sector investment levered), it is difficult to extend the same rationale to more specific and intangible sustainability criteria (quality of life, community enterprise and the social economy).
>
> (Ibid, 2004:726)

The Sustainable Urban Renewal Project Assessment Model (SURPAM)

The Sustainable Urban Renewal Project Assessment Model (SURPAM) (Lee and Chan, 2007) uses weighted indicators and data, resulting from surveys and questionnaires. Subsequent factor analysis and analytic hierarchy process can be carried out. All indicators fall under the triple bottom line. The model requires input from stakeholders and can be used to properly plan a prospective urban regeneration scheme. Citizens are able to express opinion on the design of a particular proposal before it becomes a reality with scores being expressed through multiplication of the weight of a design criterion by the score, indicating the performance of a scheme with respect to an individual criterion. However, it is cited by the authors that the model struggles to measure subjective topics such as 'sense of community', as it was impossible to develop relevant quantitative indicators for this (Lee and Chan, 2008). As a best fit, the authors solve this issue by using both cardinal and ordinal scales and Likert scales.

SURegen workbench

The SURegen workbench is an interactive online platform designed to assist those new to urban regeneration to acquire and develop skills by providing information on best practice illustrated by case studies (The University of Edinburgh, 2014). Government policies on urban regeneration aim to bring about economic, social, physical and environmental improvements in urban areas in terms of economic, social, physical and environmental improvements, however, they are often hampered by a skills-gap amongst built environment professionals, planning authorities and developers. Accordingly, this project aimed to tackle the complexity, uncertainty and ambiguity of the urban regeneration process by assisting regeneration stakeholders in day-to-day decision making and strategic processes. The workbench includes a set of integrated decision support tools to help professionals working in the field to make critical decisions, as well as to enable those who are new to the field to acquire the skills they need to meet its challenges.

The SURegen workbench addresses the issues in relation to urban regeneration. It illustrates the current scope of the UK Government's urban policies which is creating urban areas which are socially, economically and environmentally functional (The University of Edinburgh, 2014). The same government policy initiative is applied in the context of urban design. As the Department for Communities and Local Government (2014) describes good design is about creating places that work for everyone, it should look good, last well and will adopt to the needs of communities and future generations. Accordingly, this emphasises the current scope of urban design and this SURegen virtual platform can also be used by the designers in the urban design process. Chen et al. (2012) emphasise the importance of this workbench. As they stated, sustainable urban regeneration requires a comprehensive and integrated vision and action to address the resolution of urban problems and bring about a lasting improvement in the economic, physical, social and environmental conditions of an area. Accordingly, this online platform works as a decision making and knowledge sharing platform for urban planners, local authorities and other practitioners to achieve sustainability in urban regeneration activities. Furthermore, the workbench specifically addresses the management of urban regeneration projects and the skills-gap amongst regeneration professionals.

LUDA

LUDA (improving the quality of life in large urban distressed areas) is a research project which aimed to improve the quality of life in large urban distressed areas. Under the LUDA project, the programme Energy Environment and Sustainable Development within the Fifth Framework Programme of the European Union (2006) developed a handbook for a community-based approach to sustainable urban regeneration. The handbook provided insights for community-based urban regeneration and the nature of participatory workshops. The resultant project identified the need for sustainable urban regeneration as its triple bottom line (social, economic and environmental) and the need for a community-based approach in order to achieve the so-called sustainable urban regeneration. The findings of this handbook provide positive insights to develop a new community-based urban design process in order to achieve the current scope of urban design. Deakin (2009) highlights the LUDA project is an influential recent project to identify the key roles of community participation and future workshops which can play a key role in sustaining their regeneration and bringing about improvements in the quality of life.

The UK Government 'Green Book'

The UK Government's 'Green Book' focusses on techniques and procedures to follow when carrying out appraisal and evaluation and issues guidance which includes that of urban regeneration schemes. It outlines a "broad policy cycle", which can be followed by the acronym ROAMEF (Rationale, Objectives, Appraisal, Monitoring, Evaluation and Feedback), as seen in Figure 3.1.

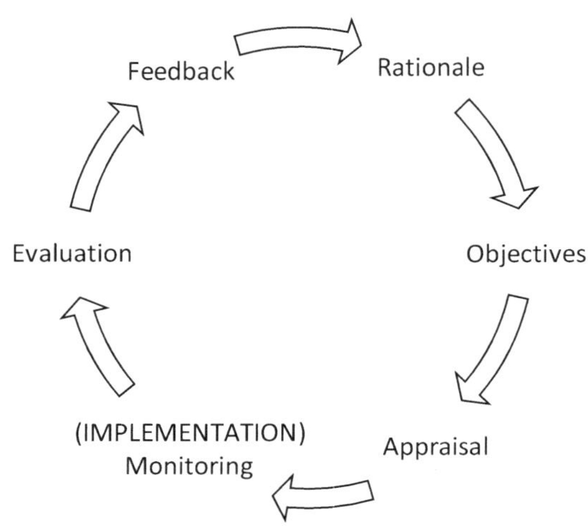

Figure 3.1 The 'ROAMEF' cycle (HM Treasury, 2018).

The evaluation process to be followed as per the Green Book guidance is to (taken directly from the Green Book Guidance, HM Treasury, 2018):

1 Establish exactly what is to be evaluated and how past outturns can be measured.
2 Choose alternative states of the world and/or alternative management decisions as counterfactuals.
3 Compare the outturn with the target outturn, and with the effects of the chosen alternative states of the world and/or management decisions.
4 Present the results and recommendations.
5 Disseminate and use the results and recommendations.

However, the Green Book's guidance on the measurement of intangible social and environmental impact relies on such techniques as social cost benefit analysis (SCBA), stated and revealed preference surveys, the subjective wellbeing approach and life satisfaction approach. Health impacts are measured using the quality adjusted life year (a measurement of disease burden).

SCBA is a form of cost benefit analysis that considers the effect of a project on environmental and social factors (Bichard, 2015). However, an issue with the SCBA approach is that it tends to only "focus on economic costs and benefits" (Ibid, 2015).

In addition, stated and revealed preference surveys are good methods of stakeholder engagement, stakeholder engagement being integral to assessment, but this engagement should be used in conjunction with other proxy-based information too, which gives a high level of scope to the numbers being used in the assessment; otherwise, despite the high level of stakeholder involvement, the certainty of the proxy could be diminished (Ibid, 2015). In addition, the subjective wellbeing and life satisfaction approaches lack a direct connection to a scheme being evaluated, which means that responses given can be vague or not directly relevant to a particular scheme.

Royal Institution of Chartered Surveyors' 'Red Book'

Of similar guidance to the 'Green Book' is the so-called 'Red Book' published by the Royal Institution of Chartered Surveyors (RICS, 2017). Although the Red Book does advise that data collection and inspection routines should be extended when working within the sustainability agenda, that sustainability considerations are considered as important when undertaking valuation assignments (Bichard, 2015) and that valuers are actively encouraged to identify and collect sustainability related data, the guidance uses the following methods of valuation which are all entirely economically based and therefore not holistic in terms of the triple bottom line of sustainable development (taken directly from the RICS 'Red Book', 2017):

- The Market Approach

  The market approach measures the value of an asset by comparing recent sales or offerings of similar or substitute property and related market data.

- The Income Approach
  The income approach measures the value of an asset by the present value of its future economic benefits. These benefits can include earnings, cost savings, tax deductions and proceeds from its disposal.
- The Cost Approach
  The cost approach indicates the value of an asset by the cost to create or replace it with another similar asset.

Summary

This chapter has provided a comprehensive review of evaluation in the context of urban regeneration. A brief history of urban regeneration evaluation in the UK context was outlined, together with a background to evaluation within urban regeneration more generally. A critical review of current assessment methodologies then followed, before the gap between such evaluation methods and the necessity for a new method was highlighted. The chapter concludes by highlighting none of the traditional and current methods of evaluation take into account the hidden social and environmental impacts of housing-led urban regeneration schemes, meaning that there is not only a niche for a quantitative evaluation approach of a scheme's socio-environmental spillovers, but that there is a necessity for a method that measures social and environmental impacts resulting from housing-led regeneration schemes in a much improved way than what has historically been before. This is imperative going forward, especially in the post Covid-19 environment.

4 Social Return on Investment (SROI)

Introduction

This chapter looks into the evaluation approach of Social Return on Investment (SROI), including its history, definition and the technique of the monetisation of intangible values, one which is integral to SROI. Further, the concept of social value is discussed, together with the recent legislation pertaining to social value, that of the Social Value Act (UK) (2012).

Further to this, the different stages of SROI are presented, together with the SROI framework and process, whilst the advantages and proclaimed disadvantages of SROI are additionally outlined. Also looked at is the notion of the SROI ratio.

Despite claims within the literature that sustainable social housing construction processes should include all three notions of the triple bottom line (Brandon and Lombardi, 2011; Langston and Ding, 2001), despite urban regeneration schemes typically failing to capture "social and environmental dimensions of creating sustainable places" (Evans, 2012), bearing in mind the difficulties experienced when attempting to capture "social and environmental gains and losses as against more easily measured economic impacts" (Ibid, 2012) and despite there being a "lack of effective tools and robust methodologies capable of capturing the full social value generated by the housing association sector" (Fujiwara and HACT, 2013), there appears to be very little within the literature base which meets these criteria and there thereby appears to be a niche for a method to be utilised which does capture such impacts – something which makes intangible social and environmental impacts tangible.

Social Return on Investment (SROI)

Social Return on Investment is considered as "the most developed method with a robust framework for implementation" (Watson and Whitley, 2016) to carry out such appraisal. It provides a framework which can help to deliver social value through built environment projects (Burke and King, 2015; Chevin, 2014; Temple et al., 2014). Indeed, SROI has been used in the context of the built environment in a number of previous studies (Aspden et al., 2012; Bichard, 2015; Bridgeman et al.,

DOI: 10.1201/9781003043881-4

2015, 2016; Watson et al., 2016; Watson and Whitley, 2016). Aspden et al. (2012) used SROI as part of a low carbon retrofit scheme in Salford, in the North West of the United Kingdom, whilst Watson and Whitley (2016) used the framework to show the impact that cancer care centres can have on building users, such as patients.

History of SROI

SROI was originally developed by George Roberts of the Roberts Enterprise Development Fund (REDF) in the United States in the mid 1990s (Watson and Whitley, 2016). The REDF was seeking methods by which they could apply commercial business acumen and practices to non-profit owned social enterprises (Emerson and Twersky, 1996). Trustees of the Fund were in need of a method to gauge the amount of social impact created and delivered from investments.

Further to SROI's initial development under the REDF, the approach was then further refined at the Harvard Business School (Maughan, 2012) and was then subsequently promoted further in the early 2000s, before a number of social enterprise practitioners then developed it further (Scholten et al., 2006).

In 2003, SROI was brought over to the United Kingdom and tested by the New Economics Foundation (NEF, 2004) before then being yet further refined by the NEF and the SROI Network in the United Kingdom (now Social Value UK; Nicholls et al., 2012). Through the work of the NEF and the SROI Network, it evolved through a new emphasis on stakeholder engagement and by using a standardised methodology (Nicholls et al., 2007). In addition, SROI sought to solve earlier problems associated with using cost-benefit analysis for sustainability evaluation through its focus on broader indicators encapsulating social, economic and environmental costs and benefits (Rotheroe and Richards, 2007).

Policy makers continue to encourage social enterprises to measure social value through the usage of SROI (Nicholls, 2007) and it continues to develop in the UK social enterprise sector as "the primary mechanism for evidencing organisational impact" (Higham et al., 2017), with endorsement from the Cabinet Office and Scottish Government (Watson and Whitley, 2016, cited in Higham et al., 2017) as well as the UK's Department of Health (UK Government, 2010) and in many other organisations such as the North West Social Value Foundation and Social Value UK in England and the Social Value Lab in Scotland (Watson et al., 2016). SROI is also being increasingly used within public policy (Musto, 2015) and commercial industry (Battle, 2014), whilst the SROI Network model has been actively promoted by the Office of the Third Sector in the United Kingdom (Zappala and Lyons, 2009).

In fact, SROI is not only taking off in the United Kingdom, but all over the world. It is now extending "beyond the US and UK as a global product" (Millar and Hall, 2013), with recent publications in Chinese and French (SROI Network, 2011), whilst the SROI Network can boast of members from all over the world (Millar and Hall, 2013).

What Is SROI

According to the SROI Network (UK), a definition of SROI is as follows (Nicholls et al., 2012):

> Social Return on Investment (SROI) is a framework for measuring and accounting for [a] much broader concept of value; it seeks to reduce inequality and environmental degradation and improve wellbeing by incorporating social, environmental and economic costs and benefits.

The SROI Network goes on to state that SROI "measures change in ways that are relevant to the people or organisations that experience or contribute to it" and "tells the story of how change is being created by measuring social, environmental and economic outcomes and uses monetary values to represent them" (Nicholls et al., 2012). Bichard (2016) adds that the SROI methodology "compares the value of the benefits of a particular action, project or programme against its costs". Semper and Fuertes (2015) add that the SROI is a "focus for understanding and managing the impact of a project, an organisation or a given policy". The authors also note SROI's traditional use within the field of project evaluation and its usefulness within the subject area of urban regeneration.

According to Scholten et al. (2006), SROI is a process of understanding, measuring and reporting on the social, environmental and economic value that is being created by an organisation and captures social value by translating social objectives into financial and non-financial measures, whilst Watson and Whitley (2016) add that SROI is a social impact methodology that allows organisations to evidence the wider value of their work and "assigns a monetary value to social returns using financial proxies, which are compared against the level of investment to produce an SROI ratio of costs to social outcomes".

Monetisation of values

The advantage that the monetising of values brings, in addition to enabling any previously intangible benefits and impacts to be measured, is that all impacts, both social and environmental, can be assessed using the same metric. Monetisation can also be seen to reduce complex information into data that can be easily compared and valued (Conejos et al., 2013; Lingane and Olsen, 2004). In addition, a novel approach is brought about through monetisation, as work carried out "to monetise the detrimental or the added sustainable value of development schemes has not featured to date in the literature" (Bichard, 2016). Indeed, Watson and Whitley (2016) add that the "main point" of difference that SROI delivers beyond existing methods in the built environment is the monetising of outcomes identified through "qualitative stakeholder engagement providing a transferable evidence base that can be communicated to a wide range of audiences". This is something that satisfies the criteria put down by Dixon et al. (2013) who suggest that there is "no universal model of urban regeneration and transferability to different contexts is often difficult".

The SROI Network Guide explains that the current approach was developed from a combination of social accounting and cost-benefit analysis and, that through an evidence-based framework, SROI "seeks to reduce inequality and environmental degradation and improve wellbeing by incorporating social, environmental and economic costs and benefits". The Guide additionally states that SROI can be carried out ex post, in an evaluative manner or ex ante, as a forecasting or predictive method (Nicholls et al., 2012).

The New Economics Foundation (NEF) outlines that SROI is "a technique widely advocated which is designed to understand, manage and report on the social, environmental and economic value created by an organisation" (NEF, 2004). The concept of 'social value' is again touched upon by Olsson and Nicholls (2005) when they outline that an SROI analysis is "a method for understanding the (environmental, social and public economic) value being created by organisations in addition to the financial value that accrues to owners".

Social value

The last two definitions of SROI above bring the notion of social value to the surface. Although there is no unequivocal definition of social value (Bichard, 2016), or even a lack of one (Ebrahim and Rangan, 2010; Polonsky and Grau, 2011), there are varying degrees of opinion. The concept of social value has, for example, been variously defined as being created when "resources, inputs, processes or policies are combined to generate improvements in the lives of individuals or society as a whole" (Emerson et al., 2001, cited in Arli and Zappala, 2009), whilst Tuan (2008) adds that social value is the "concept and practice of measuring social impacts, outcomes and outputs through the lens of cost". Emerson et al. (2001) explain that it is created "when resources, inputs, processes or policies are combined to generate improvements in the lives of individuals or society as a whole". Wood and Leighton (2010) add to this, asserting that social value refers to "wider non-financial impacts of programmes, organisations and interventions, including the wellbeing of individuals and communities, social capital and the environment".

The concept of social value has been primarily brought to the surface through the recent legislative requirements of the Social Value Act (UK) (2012).

Social Value Act (UK) (2012)

Social enterprises have come under increasing pressure to measure performance and value (Peattie and Morley, 2008) and the advent of the Public Services (Social Value) Act (2012) has further highlighted the importance, and indeed legal necessity, of evidencing social impacts of organisations and programmes. The Act requires "people who commission public services to think about how they can also secure wider social, economic and environmental benefits" and includes the notion that public service contracts take into account the wider value of a project over its entire lifetime, rather than by using a traditional cost-based procurement and "emphasises the significance of value beyond profit-driven motivations" (Watson et al., 2016).

Principles and stages of SROI

There are seven principles of SROI as follows (Nicholls et al., 2012):

- Involve stakeholders
- Understand what changes
- Value the things that matter
- Only include what is material
- Do not over-claim
- Be transparent
- Verify the result

The NEF focusses on four areas within SROI, including (Rotheroe and Richards, 2007):

- Stakeholder engagement
- Materiality
- Impact map
- Appreciation of deadweight

In addition, the various stages of SROI can be seen below (Adapted from Nicholls et al., 2012):

1 Establishing scope and identifying key stakeholders.
   Clarity is needed on exactly what the SROI will cover, who will be involved and how
2 Mapping outcomes.
   Stakeholder engagement allows the development of the impact map and creates a theory of change which shows relationships between inputs, outputs and outcomes
3 Evidencing outcomes and giving them a value.
   Researching of appropriate data to show whether outcomes have happened and subsequently valuing them
4 Establishing impact.
   Once evidence is collected on outcomes and they are subsequently monetised, aspects of change that would have happened anyway are deducted
5 Calculating the SROI.
   Calculations involve the addition of all benefits, before subtracting any negatives and comparing the result to the investment. This is also where the sensitivity of the results can be tested.
6 Reporting, using and embedding.
   This step involves sharing findings with stakeholders and responding to them, embedding good outcomes processes and verification of the report.

The above stages are contradicted slightly by the ten-stage progression outlined by the NEF (Rotheroe and Richards, 2007):

1. Understand and plan the scope of the study.
2. Stakeholders – identify and gain inputs to understand their goals and objectives.
3. Boundaries – prepare background information; learn more about the main target group.
4. Analyse income and expenditure – is financial information reported in a way that links it with economic, environmental or social objectives?
5. Impact map and indicators – understand stakeholder participation through inputs, outputs, outcomes and impacts.
6. SROI plan – summary to date. Determine the timetable for collecting the remaining data, completing calculations, report writing and sharing findings with stakeholders.
7. Implementing the plan and data collection.
8. Projections – determine for how long, if at all, projections can be justified.
9. Calculate SROI – create discounted cash flow. Use sensitivity analysis to identify areas that would yield improvements in social value.
10. Report – present results in a manner that reveals the study's subtleties, limitations and assumptions.

SROI framework and process

The SROI framework involves a robust, standardised and replicable method (Watson and Whitley, 2016) through the framework, in the form of an impact map spreadsheet, together with underlying principles and detailed guidelines and involves a high level of stakeholder interaction. It is based on traditional cost-benefit analysis and assigns monetary values to social impacts using financial proxies. These values are then compared to the level of investment to produce an SROI ratio of costs to outcomes. The framework is in the form of an impact map and includes such terms as inputs (resources that go into the intervention), outputs (product of an intervention), outcomes (difference made by an intervention), counterfactual evidence (outputs and outcomes that would have happened anyway, in spite of whether an intervention occurred or not) and impact (overall effect of outcomes, minus the counterfactual evidence) (Watson and Whitley, 2016).

There are four stages to the impact mapping process. The four stages involve the inputting of information describing:

Stage one:

- The involved stakeholders in a scheme, together with any intended or unintended changes.

Stage two:

- Inputs being invested, together with their value (in pounds); a summary of the output activities in numbers together with associated outcomes.

Stage three:

- This stage involves the associated quantifiable data, including indicators used to quantitatively measure the impact/change, the corresponding source of the indicator, the quantity in terms of numbers affected and the duration of that change. There are then further columns for the inputting of any associated proxies, together with their values and relevant sources.

Stage four:

- Stage four involves sensitivity testing calculations, including the calculation of any deadweight (what would have happened without the activity), displacement (whether another activity has been displaced), attribution (was there anyone else or any other organisation who contributed to the change) and drop off (the financial deterioration of an outcome over time [Bichard, 2015]). This stage ends in the final impact ratio for the scheme in question being created.

The SROI impact map, in spreadsheet format, can be seen in Figure 4.1.

SROI ratio

The SROI process leads to a ratio of benefits to costs being calculated. For example, a ratio of 7:1 means that an investment of £1 delivers £7 of social value (Arli and Zappala, cited in Maughan, 2012).

Within SROI parlance, money is the common unit of measurement used to show such quantifiable impact and is termed as a "widely accepted way of conveying value" (Ibid, 2012).

The SROI ratio is typically formulated as follows:

$$SROI = \frac{Net\ Present\ Value\ of\ Benefits}{Net\ Present\ Value\ of\ Investments}$$

Advantages and disadvantages of SROI

The perceived disadvantages of SROI include an over focus on financial value at the expense of a "fuller and more rounded understanding of project effects" (Durie et al., 2007), the resources required in terms of staff and time (Maughan, 2012) and abilities of built environment organisations to collect a sufficient amount of information to operate such an analysis (Hall Aitken, 2011; Trotter et al., 2014), the expense involved in the involvement of stakeholders and in carrying out "accurate economic analysis" (Maughan, 2012; NPC, 2010), which can become an inaccurate analysis, if the quantification and monetisation on the intangible social outcomes is not carried out with rigour or objectively (Krlev et al., 2013).

However, the literature cites advantages to the SROI process, including the amount of stakeholder engagement involved (Maughan, 2012) and the richness of the data collected through qualitative stakeholder engagement (Watson and Whitley, 2016), the end 'ratio', which is something commonly understood by investors, commissioners and lenders (Maughan, 2012; Watson and Whitley, 2016);

Stage 1 →		Stage 2 →			
Stakeholders	**Intended/unintended changes**	**Inputs**		**Outputs**	**The Outcomes (what**
Who do we have an affect on? Who has an effect on us?	What do you think will change for them?	What do they invest?	What is the value of the inputs in currency *(only enter numbers)*	Summary of activity in numbers	Description How would the stakeholder describe the changes?

Stage 3 →							
changes)							
Indicator	Source	Quantity	Duration	Outcomes start	Financial Proxy	Value in currency	Source
How would you measure it?	Where did you get the information from?	How much change was there?	How long does it last after end of activity? *(Only enter numbers)*	Does it start in period of activity (1) or in period after (2)	What proxy would you use to value the change?	What is the value of the change? *(Only enter numbers)*	Where did you get the information from?

Stage 4 →				
Deadweight %	**Displacement %**	**Attribution %**	**Drop off %**	**Impact**
What would have happened without the activity?	What activity did you displace?	Who else contributed to the change?	Does the outcome drop off in future years?	Quantity times financial proxy, less deadweight, displacement and attribution

Figure 4.1 Social Value United Kingdom (UK) Social Return on Investment (SROI) Impact Map (Social Value UK, 2018).

it has a usefulness in predicting or defining what is wanted from potential future contracts, in terms of social impact; it can be used in terms of strategic management – for example "monetised indicators can help management analyse what might happen if they change their strategy, as well as allowing them to evaluate the suitability of that strategy in generating social returns, or whether there may be better means of using their resources"; whilst external verification and assurance is provided through such bodies as the SROI Network (Ibid, 2012). In addition, SROI measures outcomes of interventions, rather than simply measuring outputs, with its monetisation technique facilitating "the comparison of otherwise incommensurable benefits across different activities" and allows the measurement of "what matters" to end users through a comprehensive method that is both "robust and replicable" (Watson and Whitley, 2016).

However, with existing literature so far failing to demonstrate how SROI can provide the full holistic sustainable value put forward through the notion of the triple bottom line, a "revised" form of SROI (Higham et al., 2017) has been produced by Bichard (2015). This has been termed "Sustainable Return on Investment" or "SuROI".

Summary

This chapter covered the evaluation approach of Social Return on Investment (SROI). SROI's definition, its history, the monetisation of intangible values integral to SROI, as well as the concept of social value and the concept's recent accompanying legislation – the Social Value Act (2012) – were outlined. Further to this, the different principles and stages of the SROI process were made explicit, as was its framework in the form of an impact map spreadsheet.

The SROI ratio and the advantages and disadvantages of SROI were additionally outlined. This leads onto the next chapter and an approach created from the SROI approach, that of SuROI.

5 Sustainable Return on Investment (SuROI)

Introduction

This chapter provides an understanding of the Sustainable Return on Investment (SuROI) approach to evaluation together with an in-depth explanation of the five stages involved in the calculative process, including its final ratio. There is a brief outline on ecosystem services analysis, a potential component of the environmental impact domain which makes SuROI a more sustainable framework than that of its predecessor, SROI, whilst the topic of indicators and their importance in the process is covered.

Sustainable Return on Investment (SuROI)

Langston and Ding (2001), and Brandon and Lombardi (2011), cite the importance of frameworks able to carry out full, holistic and sustainable evaluations. In addition to this, the literature shows that in real world experience, most built environment professionals working within the social housing sector are using frameworks which generate a monetary valuation of less tangible outcomes (Higham et al., 2016).

Neatly fitting into this practice, but bringing a more sustainable and holistic approach into play than is currently found within SROI (Sustainable Return on Investment; Bichard, 2015) is a structured and flexible approach to calculating the social and environmental value of change in the built environment, which can be used for this purpose. SuROI can "incorporate any approach that places a value on social and environmental change" (Bichard, 2015). The flexibility and different stakeholder and proxy values involved in a SuROI process ensure that a transferability to different contexts can be carried out, thus meeting the criteria of Dixon et al. (2013) and additionally means that business decisions related to potential built environment projects can be driven by best value in terms of sustainable impact, rather than simply by lowest cost (Higham et al., 2015).

SuROI has been described as an adaptation and extension of the SROI methodology for built environment regeneration projects (Higham et al., 2017), with the modifications allowing traditional SROI to "forecast change not only impacting on social actors, but also [on] wider aspects of economic and environmental

DOI: 10.1201/9781003043881-5

sustainability critical to the built environment and compliance with the Public Services (Social Value) Act". SROI alone has been described as failing to provide a holistic evaluation of sustainable value within the built environment (Ibid, 2017).

SuROI progresses SROI by incorporating such valuation method frameworks as Social Return on Investment (SROI), Ecosystem Services Analysis (ESA), Well-being Valuation or Life Cycle Assessment which allows the appraiser to make a quantitative valuation of both social and environmental benefits following the implementation of a regeneration scheme.

In this book, SuROI has been selected in order to: (1) deliver a quantitative assessment of both tangible and intangible benefits deriving from housing-led urban regeneration schemes; (2) add a further component to the current SuROI, thus creating a new methodology suitable to support decision making within social housing and to map stakeholders potentially interested in collaborating in social housing delivery and additionally to introduce the concept of time to the method, thus creating a management tool.

SuROI follows the process flow as set out for SROI, in the Social Value UK Guide, as found below:

1. **Establish scope and identify key stakeholders** – who will be involved in the process and how?
2. **Map outcomes** – through engagement with stakeholders, an impact map, or theory of change, will be developed which shows the relationship between inputs, outputs and outcomes.
3. **Evidence outcomes and give them a value** – this stage involves finding data to show whether outcomes have happened and then valuing them.
4. **Establish impact** – having collected evidence on outcomes and monetised them, those aspects of change that would have happened anyway or are as a result of other factors are eliminated from consideration.
5. **Calculate** – this stage involves adding up all the benefits, subtracting any negatives and comparing the result to the investment.
6. **Report, use and embed** – the last step involves sharing findings with stakeholders and responding to them, embedding good outcome processes and verification of the report.

According to Nicholls et al. (2012:8), "things which get bought and sold take on greater significance". SuROI takes this mantra on board (Bichard, 2015) and places a numerical value on the social and environmental components of the triple bottom line; aspects which are normally difficult to compare against more easily measured economic impacts (Conejos et al., 2013). Other authors cite the typical difficulty in measuring the social 'pillar' due to its abstract nature and consider that it could be overshadowed by the economic and environmental domains (Davidson, 2009; Lehtonen, 2004; Littig and Griessler, 2005). Xing et al. (2009:210) cite that one of the main challenges can be a difficulty in the measurement of what they term "apples and pears", which is to compare the measurement of costs and values which are expressed in different units. The SuROI method potentially solves this issue. SuROI also satisfies evaluation criteria set out by Roberts (2006) who cites that an evaluation should be inclusive, open, accountable and transparent, that it should take into account the views and objectives of all actors involved, that the methodology in question should allow meaningful comparisons to occur, that the methodology can be adjusted to fit in with local or regional circumstances and that it can be used to enhance participation and engagement with different actors and organisations. Similarly, it doesn't just take into account the "standard indicator which pays little attention to prevailing conditions" but does involve "tailored to fit" indicators (Ibid, 2006) and proxy values.

In addition, recent guidance from the Royal Institution of Chartered Surveyors (2014) recognises the need to include a wider range of factors that can influence the value of built environment projects and asserts that sustainability considerations are now considered as being important when undertaking valuations (Bichard, 2015). The aforementioned Social Value Act (2012) additionally requires that economic, environmental and social benefits are taken into consideration as part of any procurement processes, showing that the focus on sustainability is perhaps starting to change and is potentially becoming more important at national level also.

SuROI aims to allow the environmental and social value of a project, programme or policy in the built environment to be made explicit through evidence and be added to capital costs to give an overall sustainable value (Bichard, 2015). As has been seen previously, SROI compares the prospective social benefits of a particular scheme against its costs and ESA (ecosystem services analysis) takes the costs and benefits of the environment into consideration. ESA covers both the natural and built environment, including architectural aspects within its definition (DEFRA, 2007).

SuROI is heavily influenced by SROI and incorporates the same Social Value UK impact map (Figure 6.2) that SROI utilises as its framework (Bichard, 2015). Although SROI takes into account only social value, amendments to SROI by Bichard (2015) outline how environmental value can also be taken into account and such quantifiable impact inputted into the impact map framework format, thus providing a more sustainable impact calculation than SROI otherwise would.

SROI measures social change and uses monetary values to represent that change (Nicholls et al., 2012) by comparing the value of the benefits of a project against its costs. SROI is flexible (other approaches can be incorporated into the impact map shown below) and its principles ensure a consistency in approach. It is the flexibility of the SROI methodology which lends itself to an incorporation of additional approaches. The practical workings of the SROI methodology involve the usage of the impact map in the form of a Microsoft Excel spreadsheet. SROI uses the impact map to list and subsequently monetise social changes, benefits and impacts (both positive and negative) and to show all calculations leading to a cost/benefit impact ratio (Nicholls et al., 2012). Information populated includes stakeholders, inputs, outputs (number of units of delivery) and outcomes (stakeholder-defined change). Because the impact map is able to incorporate other approaches into its framework, this has led to environmental impacts being able to be measured, this being additional to the previously measured social value, within the parameter of the impact map. In the same way that social value can be measured through SROI or Wellbeing Valuation (measurement of life satisfaction), environmental value can be measured through Ecosystem Services Analysis (ESA) (costs and benefits of the natural environment). All values can be inputted, costed and calculated within the same impact map.

Ecosystem services analysis

Social and ecological systems are closely linked (Folke, 2006; Young et al., 2006), with the world's ecosystems sometimes being referred to as capital assets which, if properly managed, can yield a flow of vital services (Daily et al., 2000).

Ecological systems contribute in many ways. They are of great importance to human survival (TEEB, 2011), are highly valuable – the cost of ecosystem services in 1997 estimated by Costanza et al. (1997) would have been nearly twice the entire gross national product at that time. In addition, if one ecosystem service were to decline, this would impact on future production capabilities in another (Shelton et al., 2001).

In addition, natural ecosystems play an essential role in the regulation and maintenance of ecological processes and life support systems on earth. They include such aspects as gas regulation, climate regulation, disturbance prevention, water regulation, water supply, soil retention, soil formation, nutrient cycling, waste treatment, pollination, biological control, refugium function, nursery function, food, raw materials, genetic resources, medicinal resources, ornamental resources, aesthetics, recreation, cultural and artistic inspiration, spiritual and historical elements and scientific and educational information (De Groot et al., 2002). A definition of ecosystem services by the same author in 1992 asserts that ecosystem functions as the "capacity of natural processes and components to provide goods and services that satisfy human needs, directly or indirectly" (De Groot, 1992).

Historically, there have been references to ecosystem services and their economic value dating back to the 1960s and 1970s (Helliwell, 1969; King, 1966; Odum and Odum, 1972). And in recent times, there has been a growth in publications on the benefits of ecosystems to human society (Bingham et al., 1995; Costanza et al., 1997; Daily, 1997; Daily et al., 2000; De Groot, 1992, 1994; Pearce, 1993; Turner, 1993; Wilson and Carpenter, 1999).

It has been regarded as being difficult to understand the complexity and 'public goods' nature of ecosystems (Costanza et al., 1997) and has also been cited that gaps exist in terms of the measurement of ecosystem services (ten Brink et al., 2011). However, contrary to this, it is important to cite that many databases are available (such as the TEEB, the Economics of Ecosystems and Biodiversity Valuation, database) to provide a large amount of indicators and proxy values for ecosystem services, whilst "each ecosystem service has aspects that can be measured" (Zari, 2012). It is also important to note that as human ingenuity in deriving benefit from ecosystem services and lists of known ecosystem services evolve and expand over time, it is likely that more will come about in the future (Heal et al., 2005).

De Groot et al. (2000) group ecosystem functions into four primary categories:

1 Regulation functions

This relates to the capacity of natural and semi-natural ecosystems to regulate essential ecological processes. (Benefits to humans from regulation include clean air, water, soil and biological control services).

2 Habitat functions

Natural habitats provide refuge and reproduction habitat to plants and animals which enables the species to continue, to continue to evolve and the conservation of such species to continue.

3 Production functions

Photosynthesis and nutrient uptake converts energy, carbon dioxide, water and nutrients into carbohydrate structures, used by secondary producers to create a further and larger variety of living biomass. Human services resulting from this include food, raw materials, energy resources and genetic material.

4 Information functions

Natural ecosystems provide a reference function and contribute to the maintenance of human health through opportunities for reflection, spiritual enrichment, cognitive development, recreation and aesthetic experience (Table 5.1).

SuROI process and stages

The SuROI approach assesses the degree to which change has occurred, both in terms of the significance of the change and the number of people experiencing that change. Both social and environmental outcomes are then able to be monetised by identifying an appropriate indicator for the change and then by applying a suitable monetary value or proxy to each indicator. Indicators and proxies are typically taken from wellbeing sources or socio-economic statistics compiled by a range of agencies. The values are multiplied by the numbers affected and the amount of time the change is expected to last for (Bichard, 2016). SuROI combines stakeholder accounts and statistical trends as part of its calculations, to ensure "a robust and defendable result" (Ibid, 2016). As previously described, because all value is monetised, this has the advantage of measuring and presenting information according to the same metric. The social monetary changes are added to the environmental monetary changes to give a gross value used to calculate SuROI. The net value is then subsequently derived once adjustments are made, including deadweight (the amount of outcome that would have occurred anyway), displacement (the amount of activity that has been displaced), attribution (the amount of outcome caused by other interventions) and drop off (the deterioration of an outcome over time).

Table 5.1 Shows ecosystem services relating to the built environment

Ecosystem service	*Ranking criteria*			*Examples of existing design methods that could potentially be used*	*Positive environmental implications*
	Applicability to the built environment	*Ecological significance*	*Negative environmental impact caused by the built environment*		
Supporting services					
Habitat provision (including provision of genetic information; biological; fixation of solar energy; and species maintenance)	Medium	High	High at a local scale	Revegetation; preservation of existing flora and fauna; urban wildlife sanctuaries; living walls; urban forests; green roofs and facades; wildlife corridors; green belts	Increased biodiversity; reduction of the urban heat island effect; sequestration of carbon; increased air, water and soil quality; possible protection from wind or wave surges; more adaptable ecosystems as the climate changes; reduction of storm water peak flows
Nutrient cycling (including: decomposition; soil building; and the provision of raw materials)	Medium	High	High at regional/ global scale	Recycling and reuse techniques; cradle-to-cradle design; composting techniques; design for deconstruction; landfill mining' industrial ecology	Reduction of waste; reduced need for mining/growing/production/ transportation of materials and energy leading to reduction in greenhouse gas (GHG) emissions, waste and ecosystem disturbance; decreased use of energy; increased health of ecosystems and humans
Regulation services					
Purification	High	High	High at regional/ global scale	Living machines; phyto-remediation and bio-remediation; filtration techniques; green roofs and facades; urban forests; constructed wetlands; composting techniques	Increased health of living organisms; increased terrestrial and marine productivity; reduction of air and water pollution; eutrophication reduction; remediation of polluted sites; reduced ozone damaging gas and GHG emissions

Table 5.1 (Continued)

Ecosystem service	*Ranking criteria*			*Examples of existing design methods that could potentially be used*	*Positive environmental implications*
	Applicability to the built environment	*Ecological significance*	*Negative environmental impact caused by the built environment*		
Climate regulation	High	High	High at a global scale	Storage of carbon in building structure; revegetation; design to enable behavior change in energy use; renewable energy generation; passive solar design; non-high thermal mass infrastructure and landscaping; design to reduce reliance on fossil fuels	Mitigation of the causes of climate change; more adaptable communities; mitigation of the living heat island effect; improved health of living organisms
Provisioning services					
Provision of fuel/energy for human consumption	High	High	High at a global scale	Design for renewable energy generation; co-generation methods; design to enable behaviour change to reduce energy use; industrial/construction ecology	Reduced transport and energy generation-related GHG emissions; more self-reliant and therefore robust urban environments; reduction of air, water and soil pollution; reduction of mining and drilling impacts
Provision of fresh water	High	High	High at a regional scale	Rain water harvesting and storage; grey/black water recycling; design incorporating water saving equipment; porous paving surfaces; water efficient landscaping	Reduction of water pollution; increased health of riparian systems; reduction of the urban heat island effect; increased quality of water; increased health of living organisms

Data used within the impact map can be that of primary data, secondary data or both. Methods of primary data include interviews, focus groups, surveys or questionnaires, whilst secondary data includes data sets such as social/economic statistical sets, indicators, proxies and/or databases. In order to estimate the value of non-traded and non-market goods, financial proxies are used.

An example of the outline of the impact map can be seen in Figure 5.1. Stakeholder data collected in the field can be used to evaluate social and environmental change caused by previously completed projects, and also to predict future change from planned projects.

The impact map requires the inputting of details and figures into its spreadsheet format to ultimately arrive at the final quantitative impact ratio of a project. Such details include the establishment of stakeholders (those who influence the project), inputs (costs of the project), outputs (number of units of delivery) and outcomes (predicted/stakeholder change) (Higham et al., 2017). The approach attributes financial value to inputs and outcomes (Arvidson et al., 2013).

There are five stages involved in a SuROI calculation. Four stages can be seen within the parameter of the impact map, in spreadsheet format, with the final stage 5 involving the final ratio-centred calculations:

- **Stage One:**

  Stage one of the process involves the establishment of a project's scope, making explicit the aims and objectives of a project, including "features of the scheme that it is appropriate to measure" (Higham et al., 2017) and intended/unintended changes arising from a project (Nicholls et al., 2012). The identification of relevant and affected stakeholders who have an effect on a project and/or who are affected by it is additionally carried out. Nicholls et al. (2012) and Bichard (2015) cite the importance of involving stakeholders at this stage.
- **Stage Two:**

  On completion of stage one, stage two of the impact map involves the identification of impacts, the valuing of inputs and the clarification and description of outputs involved. Nicholls et al. (2012) cite that the process involves the establishing of inputs which then subsequently lead to delivering activities (outputs) which then result in impact or change (outcomes) for identified stakeholders.
- **Stage Three:**

  Stage three involves the evidencing of what changes. This stage involves fieldwork and desk-based research to establish outcomes and their relevant monetary values (Watson and Whitley, 2016). At this stage, interviews, focus groups, questionnaires or workshops can be held with involved stakeholders.

  Outcomes are determined and their effects quantified, with monetary values attached to each outcome. Such monetary values can come in the form of financial proxies (Bichard, 2015) which can be gleaned from primary data, or academic, public or social enterprise literatures (Higham et al., 2017), or existing social value assessments or datasets such as the HACT database of Wellbeing Valuation (Trotter et al., 2014) or the Global Value Exchange (GVE, 2017). The "total incidence of impact multiplied by the proxy determines the value created by each outcome for a specific user group in a single year" (Higham et al., 2017). The duration of an impact is also highlighted within this stage.

- **Stage Four:**
  Stage four involves the establishment of the impact and involves taking into account counterfactuals such as deadweight and displacement, attribution, drop off and finally the calculation of the impact (Nicholls et al., 2012). This enhances the validity of the calculations and impact (Bichard, 2015) and "allows adjustments to be made to the initial valuation ensuring it provides a reasonable representation of new impact" (Higham et al., 2017).
- **Stage Five:**
  Stage five involves the calculation of the SuROI ratio, whilst also projecting impact into the future and additionally undertaking a sensitivity analysis (Nicholls et al., 2012), designed to "mitigate the effects of risk" (Higham et al., 2017).

The SuROI ratio is calculated by dividing the present value over the value of inputs. The ratio does not express financial value, but is instead a comprehensive way of expressing the "currency of social value" (Arvidson et al., 2013). The future projection of drop off over time is calculated as part of the impact map spreadsheet and discounts using the 3.5% social time preference rate outlined in HM Treasury's *Green Book* (HM Treasury, 2018). The Sensitivity Analysis is carried out to ensure validity (Bichard, 2015; Nicholls et al., 2012). RICS (the Royal Institution of Charted Surveyors) guidance on development appraisal recommends that any financial appraisal has full counterfactual and sensitivity testing prior to reporting (RICS, 2012). Sensitivity checks can involve estimates of deadweight, attribution, drop off, financial proxies, quantity of outcomes and values of inputs (Higham et al., 2017).

Sensitivity testing aims to calculate the amount each counterfactual needs to be changed to take the SuROI ratio to 1.0, i.e. from positive to negative, or vice-versa, which guards against potential over claiming (Nicholls et al., 2012).

One of the main attributes to SuROI, as stated previously, is the quantitative measurement of previously intangible social and environmental spillovers. It is important that hidden social and environmental benefits are taken into consideration upon carrying out an evaluation for the following reasons:

- To present a basic knowledge of what a scheme can offer in terms of social and environmental impact;
- To ensure that financial decisions are being made with the maximum amount of information being available on social and environmental benefits;
- Housing-led urban regeneration schemes being evaluated as "successful" schemes might not be "successful" and vice versa, which would lead to incorrect decision making in terms of future targeted investment, and potential wasting of money as a consequence;

Stage 1 →		Stage 2 →			
Stakeholders	**Intended/unintended changes**	**Inputs**		**Outputs**	**The Outcomes (what**
Who do we have an affect on? Who has an effect on us?	What do you think will change for them?	What do they invest?	What is the value of the inputs in currency *(only enter numbers)*	Summary of activity in numbers	Description How would the stakeholder describe the changes?

Stage 3 →							
changes)							
Indicator	Source	Quantity	Duration	Outcomes start	Financial Proxy	Value in currency	Source
How would you measure it?	Where did you get the information from?	How much change was there?	How long does it last after end of activity? (*Only enter numbers*)	Does it start in period of activity (1) or in period after (2)	What proxy would you use to value the change?	What is the value of the change? (*Only enter numbers*)	Where did you get the information from?

Stage 4 →				
Deadweight %	**Displacement %**	**Attribution %**	**Drop off %**	**Impact**
What would have happened without the activity?	What activity did you displace?	Who else contributed to the change?	Does the outcome drop off in future years?	Quantity times financial proxy, less deadweight, displacement and attribution

Figure 5.1 The SuROI impact map, which uses the Social Value UK SROI impact map as its framework (Social Value UK, 2018).

- Social and environmental benefits "of central concern to individuals and communities" (Vardakoulias, 2013) are potentially not being highlighted and measured correctly, which is potentially having a direct impact on residents' quality of life.

General issues and issues with indicators

A SuROI analysis is usually formed from data received as a combination of first-hand primary data from involved and related stakeholders, in combination with secondary data, in the form of indicators and proxies (Bichard, 2015, 2016). However, an analysis is only as good as the data that goes into it and which formulates it. Some issues to bear in mind when a SuROI analysis is performed includes the importance of choosing datasets, indicators and financial proxies which are specific, reliable, accurate and applicable (Higham et al., 2017).

Additionally, as is the case with SROI, it can be argued that there is significant time and expense in carrying out such an analysis, both in terms of collecting the data and in terms of the associated calculations. However this author's first-hand experience, together with feedback received as part of the interviews carried out later on within this book with practitioners in the field, would suggest that the costs incurred through the usage of previous evaluation methods nevertheless promote great enthusiasm with a way forward such as the approach of SuROI.

There has additionally been an argument within the literature base that there is an inability to benchmark SuROI valuations. Because of the flexible nature of the SuROI approach, namely that the various interrelated inputs, outputs and outcomes for different schemes can all be entered into the SuROI impact map, this does mean that an absolute direct benchmarking is impossible, simply because all schemes are different. This is, however, the case with any evaluation methodology. What can be stated is that SuROI presents more of a 'benchmarkability' with regard to the measurement of previously intangible social and environmental impact, because previously this has never before been carried out in such a quantifiable and holistic fashion.

Summary

This chapter provided an understanding of the Sustainable Return on Investment (SuROI) approach to evaluation. SuROI's origins stem from SROI, however, SuROI incorporates an environmental impact to its evaluation ratio, meaning that it provides a more sustainable assessment than by simply looking at social impacts. There was an in-depth explanation of the five stages involved in the calculative process, including the final SuROI ratio. There was a brief outline on ecosystem services analysis, a potential component of the environmental impact domain which makes SuROI a more sustainable framework than that of its predecessor, SROI, whilst the topic of indicators and their importance in the process was covered. The next chapter will describe SuHousingImpact, which bridges the gaps between SROI and SuROI.

6 An introduction to a new stakeholder-led evaluation method

SuHousingImpact

Introduction

This chapter will explain exactly how a new stakeholder-led evaluation method – the 'SuHousingImpact' tool which has been developed – functions. Gaps within SuROI are explained, whilst the Multi-Stakeholder Partnership approach, Multi-Stakeholder Networks, Stakeholder Management and the subject area of Stakeholder Analysis, all areas which link in to the new evaluation tool, will be covered, together with the theoretical underpinning of the method and how the tool relates directly to common budgetary issues found within the modern day New Public Management structures in place.

A new evaluation method

Gaps within the SuROI methodology

The SuROI methodology incorporates in its very essence the usage of stakeholder involvement. Subsequently, gaps within the domain of stakeholder analysis can be highlighted, whilst gaps within the SuROI literature base can also be made explicit, to better clarify the formation of the new SuROI refined method that this book subsequently focuses on.

Cross co-operation across and amongst different stakeholders is important when attempting to reverse the effects of urban degeneration of areas and when attempting to promote socio-environmental impacts to an area. The cross co-operation amongst stakeholders promoted as part of the method of this book can be a particularly useful concept with regard to public policies, theoretically and with specific reference to housing.

Public policies

Public policy decision making has been referred to as almost always involving "multi-party decisions", with both procedural and communication components involved amongst parties which affect outcomes (Fischer et al., 2007). Indeed, Andrews (2007) cites the example regularly seen as the norm in society, typically

DOI: 10.1201/9781003043881-6

at a local authority level, of a lone person within a public domain making decisions affecting an entire community. The author adds that it is important that we move away from this, towards a more involved and stakeholder-oriented approach, thus again tallying with the method within this research.

The above approach can be referred to as a Multi-Stakeholder Partnership (MSP) approach where groups of individuals are brought together to collectively drive a process that seeks to change a common environment (Truex and Soreide, 2010) and which can potentially create a "new form of global governance with the potential to bridge multilateral norms and local action by drawing on a diverse number of actors in civil society, government and business" (Bäckstrand, 2005). Partnerships have been referred to as implementation networks which have the potential to "bridge global multilateral norms and local action in areas such as sustainable development" (Joyner, 2005), which can potentially "operationalise lofty principles of sustainable development" (Bäckstrand, 2005). Indeed, more than 200 such partnerships for sustainable development purposes were announced as part of the World Summit on Sustainable Development held in Johannesburg in 2002 (Ibid, 2005). Deepak (2011) concurs, adding that MSPs are "unique in both character and substance" and are generally directed at the problems and challenges of sustainable development.

MSPs are increasingly important mechanisms of improving corporate social responsibility (Seitanidi and Crane, 2009), solving social problems (Kanter, 1999), improving social welfare (Berger et al., 2004) and achieving strategic corporate goals (Sloan, 2009).

The emergence of MSPs within the field of sustainable development, particularly useful as guidance and background within this book, includes "the emergence of governance structures based on private authority, private regimes and a mix of public and private actors" and adds that such partnerships have been brought about to create solutions where "intergovernmental diplomacy alone cannot grapple with the pressing problems and complex dimensions of sustainable development" (Bäckstrand, 2005).

MSPs have been referred to as being

> about sharing not shifting risks; finding innovative ways to pool resources and talents based on each party's core strengths; and designed and maintained over time in such a way as to deliver mutual benefits for all collaborating parties, whilst working together in such a way as to promote shared visions, joint problem solving, a work ethos that exploits mutual self-interest and adds value.
>
> (Deepak, 2011)

The partnerships involved in MSPs have been described as (Deepak, 2011):

1 A means to an end and not an end in themselves (they are formed to actually achieve something).
2 A sharing of resources and responsibility (resources not necessarily only involving money, but knowledge and skills too).

3 Needing to give up some element of their own control (which carries both risks and benefits).
4 Having a sufficient amount of mutual respect and trust between partners to enable them to work together and sufficient overlap of goals.

The set of principles for MSPs drawn up as preparatory input to the World Summit on Sustainable Development in 2002 entitled the "Bali Principles" cite that partnerships should (GKP, 2003):

1 Help achieve the further implementation of Agenda 21 and Millennium Development Goals, consistent with sustainable development strategies and poverty reduction strategies;
2 Be voluntary and self-organising;
3 Be based on mutual respect and shared responsibility of the partners involved;
4 Have a multi-stakeholder approach, arranged among any combination of partners, including governments, regional groups, local authorities, non-governmental actors, international institutions and private sector partners; and
5 Be international in their impact, beyond the national level.

Key operational challenges of effective partnerships have been cited in the literature. Within Malena's (2004) background paper for a multi-stakeholder workshop on partnerships and UN-Civil Society relations in New York in 2004 which described different categories pertaining to the process involved in MSPs, it is cited that the following aspects are of importance:

- **Inclusion:**
  As Malena states "a first key operational challenge of an effective partnership is getting the right actors around the table". It is cited that stakeholder analyses, the placing of particular emphasis on primary stakeholders, the development of a strategy or action plan to promote inclusion and the use of clear and transparent processes to identify potential partners are all useful strategies.
- **Clear definition of purpose and roles:**
  It is important to ensure that purpose and expected results of a partnership, together with the respective roles and responsibilities of each partner are clearly defined and commonly agreed. A common purpose should be defined, together with the establishment of accepted parameters of divergence, a focus on results and a clear negotiation of partner roles. Additionally, specific commitments should be encouraged.
- **Participation/power sharing:**
  This category involves the challenge of ensuring effective participation and appropriate power sharing within MSPs. Guidance is provided by suggesting an open acknowledgement and addressing of power differentials, the establishment of appropriate decision making structures and rules and possibly even toolkits utilised to facilitate decision making or conflict resolution.

- **Accountability:**
  Included within the category of accountability are such methods as the drafting of an accountability map and strategy, transparency and communication, emphasising and supporting links between partners and the development and adherence to strict performance standards, monitoring and reporting requirements.
- **Strategic influence:**
  The maximisation of strategic influence can be achieved through such methods as the cementing of partnerships to agreed priorities and targets, the development of a strategy relating to official agendas and processes and ensuring that processes are strategically anchored within host organisations and the setting of strategic as well as organisational goals.
- **Stakeholder management:**
  Stakeholder management has been cited as being a vital part of any project deliverables (Chang et al., 2013), whilst researchers elsewhere have described stakeholder management as a "systematic approach to organise the relationship between business and society" (Roloff, 2008), or a process in which a project team facilitates the encompassing needs of stakeholders to identify, discuss, agree or contribute to achieve their objectives (Brammer and Millington, 2004; Pajunen, 2006; Rowlinson and Cheung, 2008). Kerzner (2011:34) cites six processes being involved in stakeholder management, including "identifying stakeholders, analysing, engaging, identifying information flow, enforcing stakeholder agreement and stakeholder debriefing", whilst contrary to this, Eskerod and Jepsen (2013) assert that there are only the three processes involved – those of stakeholder identification, assessment and prioritisation. There are typically two categories of stakeholder management, one of organisational-focussed stakeholder management and one of issue-focussed stakeholder management. Issue-focussed stakeholder management is the category most pertinent to this research and involves the building of social capital. It has the "potential to initiate more sustainable solutions" (Roloff, 2008). The differences between the two concepts are outlined within Table 6.1.

Stakeholder management has been referred to as a "fundamentally pragmatic concept" (Freeman, 1999: 234), thus fitting in with the pragmatic nature of this research.

- **Multi-stakeholder networks:**
  As Roloff (2008) cites, "whenever a problem or challenge relates to several stakeholders, it is likely to have a complexity that cannot be handled by one actor. In this situation the companies involved should adapt an issue-focussed stakeholder management approach that aims to solve the problem by communication and collaboration between the stakeholders of the issue".

  Multi-stakeholder networks are networks comprised from "business, civil society and governmental or supranational institutions" who come together

Table 6.1 The characteristics of stakeholder management approaches focusing on organisations or issues (taken from Roloff, 2008)

Focus on	*Organisation*	*Issue*
Point of reference	Focal organisation, e.g. a corporation	A mutual problem, a crisis, a conflict, a challenge
Stakeholder definition	A group who can affect or is affected	A group who is affected by the solution
Aim	Protect the focal organisation	Approach an issue, solve a problem
Method	Relationship management	Identification and implementation of a common course of action
Initiator/organiser	Focal organisation	Stakeholder or focal organisation
Topics discussed	The conduct and misconduct of the focal organisation	Structures that cause or enforce the problem/crises/conflict
Means of communication	Reports, press releases, public discussions aiming at potential customers, personal interactions, etc.	Face-to-face interactions in groups, inclusion of new participants, but exclusion of onlookers, personal interactions, etc.
Mode of interaction	Strategic action	Communicative action
Type of legitimacy	Pragmatic legitimacy	Moral legitimacy

to find a "common approach to an issue that affects them all and that is too complex to be addressed effectively without collaboration" and indeed often forces stakeholders on the issue to co-operate with each other (Roloff, 2008). Many authors refer to multi-stakeholder networks as functioning as an additional venue for democratic participation (Hajer and Wagenaar, 2003; Rhodes, 2000).

Roloff (2008) describes there being seven phases "that amount to a life cycle" within typical processes encountered by Multi-Stakeholder Networks – initiation, acquaintance, first agreement, second agreement, implementation, consolidation and institutionalisation or extinction:

- **Initiation phase:**
  This phase consists of the period of time when Multi-Stakeholder Networks emerge, as a problem or a challenge becomes urgent for several different actors.
- **Acquaintance phase:**
  The acquaintance phase comprises of understanding the opinions of other involved stakeholders, together with each stakeholder's separate interpretation of the problem. Participants also begin to grasp the complexity of the issue, learn about interdependencies that were not apparent before and learn to respect each other's concerns.
- **First agreement phase:**
  The first agreement phase involves different accounts of the situation being presented and discussed with the aim of agreeing upon a common description.

This ideally involves an analysis of issue-related knowledge and shared interpretations of this knowledge to ultimately lead to a problem definition (Rauschmayer and Wittmer, 2006) or to shared objectives.

- **Second agreement phase:**
  The second agreement phase is used to compare the different approaches available and to subsequently select one or more for implementation.
- **Implementation phase:**
  The implementation phase involves the proposed approach being broken down into activities and a division of labour as agreed upon by the stakeholders involved, with every stakeholder honouring his or her promises and contributing to the mutual aim.
- **Consolidation phase:**
  In this phase, the experience of co-operation between stakeholders should have built up some levels of trust. Activities and procedures are then institutionalised with potential constitutions, rules, divisions of work and contributions between stakeholders being written if applicable. New stakeholders typically join the network in this phase when abilities are proven and rules for participation and co-operation are well established.
- **Institutionalisation/extinction phase:**
  The final phase involves the network of stakeholders either becoming established and carrying out and solving tasks according to earlier phases, or alternatively becoming extinct.

In addition to the above process models, Table 6.2 cites some alternative viewpoints from the literature, with a focus on stakeholder management process models within construction projects in the built environment; of relevance to this domain.

The Kaldor-Hicks compensation criterion

The field of microeconomics, and more specifically, the Kaldor-Hicks compensation criterion, acts as a theoretical basis for the offloading of costs onto stakeholders described in this book, and acts as a fundamental criterion of the SuHousingImpact method.

Public policy decision making has been referred to as almost always involving "multi-party decisions", with both procedural and communication components involved amongst parties affecting outcomes (Fischer et al., 2007). Indeed, Andrews (2007) cites the example previously and typically used in society of a lone person within a public domain making decisions affecting an entire community, adding that it is important that we move away from this, towards a more involved approach. It is the science behind stakeholder co-operation which serves a fundamental component of the method created within this book and which relates to both microeconomics, welfare economics and the 'compensation principle' which states that if prospective gainers could compensate prospective losers and leave no one worse off, the alternate state is to be selected (Chipman, 1987).

Table 6.2 Stakeholder management process models in construction projects according to the literature (taken from Yang et al., 2009)

Scholars	*Stakeholder management processes*
Karlsen (2002)	Identification of stakeholders; analysing the characteristics of stakeholders; communicating and sharing information with stakeholders; developing strategies, following up.
Elias et al. (2002)	Developing a stakeholder map of the project; preparing a chart of specific stakeholders; identifying the stakes of stakeholders; preparing a power versus stake grid; conducting a process-level stakeholder analysis; conducting a transactional-level stakeholder analysis; determining the stakeholder management capability of the projects; analysing the dynamics of stakeholder interactions.
Young (2006)	Identifying stakeholders; gathering information about stakeholders; analysing the influence of stakeholders.
Bourne and Walker (2006)	Identifying stakeholders; prioritising stakeholders; developing a stakeholder engagement strategy.
Olander (2006)	Identification of stakeholders; gathering information on stakeholders; identifying stakeholder mission; determining stakeholder strengths and weaknesses; identifying stakeholder strategy; predicting stakeholder behaviour; implementing stakeholder management strategy.
Walker et al. (2008)	Identifying stakeholders; prioritising stakeholders; visualising stakeholders; engaging stakeholders; monitoring effectiveness of communication.
Jepsen and Eskerod (2009)	Identification of the (important) stakeholders; characterisation of the stakeholders pointing out their (a) needed contributions, (b) expectations concerning rewards for contributions, (c) power in relation to the project; decision about which strategy.

In particular, it is the Kaldor-Hicks compensation criterion which originates from Nicholas Kaldor (1939) and John Hicks (1939) which proves particularly useful. Kaldor and Hicks describe the criterion at the time of their writing, through citing the example of the English Corn Laws in 1846 which "removed legal protections of corn farming" (Bostani and Malekpoor, 2012). This harmed landowners, but helped consumers of bread. Kaldor and Hicks state that if those gaining could theoretically compensate those harmed and yet still remain better off, then this is a desirable situation. Note that compensation potentially paid is a theoretical possibility rather than a factual certainty. The Kaldor-Hicks criterion can be contrasted to the Pareto Principle (also known as the 80/20 rule, the law of the vital few, or the principle of factor sparsity; states that, for many events, roughly 80% of the effects come from 20% of the causes) which states that there must be no losers in the same circumstances.

The key requirement for a Kaldor-Hicks analysis is the "ability to gather and sum up people's willingness to pay attached to different outcomes" (Stringham, 2001). Kaldor-Hicks efficiency criterion implies that beneficiaries of a decision should hypothetically be able to compensate losses of losers. The measurement unit cited to compensate potential loss is not utility, but rather money (Bostani and Malekpoor, 2012).

According to Deepak (2011), co-operation between all stakeholders is fundamental to ensure the sustainability of programmes. Indeed, the author cites that an unequal participatory process can actually endanger such sustainability due to the lack of participation involved, before adding that co-operation and co-ordination between all sectors within society is imperative to ensure an efficient social housing sector.

The new SuHousingImpact method refines SuROI. One of the refinements includes the offloading of costs onto potential stakeholders by combining the usefulness and practicalities of stakeholder networks with the theoretical underpinning of the Kaldor-Hicks criterion.

By implementing SuROI in the housing sector, it should be easier to activate additional stakeholders whose benefits in participating and supporting housing programmes was previously not perceived, while after quantifying them, they could be more willing to invest having seen through an evidence-based methodology, the positive spillovers (or the saved costs) they would incur in the case of non-action.

Stakeholders analysis: limitations and potential for honing the SuROI methodology

A stakeholder analysis can be used to inform project planning, implementation or evaluation (MacArthur, 1997; ODA, 1995) by typically identifying key persons involved in a project, and subsequently assessing their knowledge, interests, positions, alliances and importance related to the policy in question. This enables more interaction, success and the prevention of potential problems (Schmeer, 1999). A stakeholder approach reflects the realisation that the interests and influence of individuals or groups, both within and outside a prospective organisation, need to be taken into consideration (Brugha and Varvasovszky, 2000).

Stakeholder analysis was derived from business management (Reed et al., 2009), with the original concept behind stakeholder analysis being to identify any stakeholder groups wishing to threaten corporate organisations. This changed full swing over time, to give a voice to marginalised stakeholders in order that decision making by the powerful could be influenced.

Stakeholders definitions

Stakeholders are integral to a stakeholder analysis. Table 6.3 provides some definitions of the term 'stakeholder'.

Definition of stakeholder analysis

There are many references to the definition of the term 'stakeholder analysis' in the literature. Common keywords appear in the definitions. The notion of

Table 6.3 Definitions of the term 'stakeholder' within the literature

Author	*Definition of 'stakeholder'*
Freeman (1984)	"any group or individual who can affect or who is affected by the achievement of the organisation's objectives"
Bryson (1995)	"any person, group or organisation that can place a claim on the organisation's attention, resources or output, or is affected by that output"
Morgan and Taschereau (1996)	"as persons or groups whose interests and activities strongly affect and are affected by the issues concerned, who have a stake in a change, who control relevant information and resources and whose support is needed in order to implement the change"
Brugha and Varvasovszky (2000)	"individuals, groups, and organisations who have an interest (stake) and the potential to influence the actions and aims of an organisation, project, or policy direction"
Jonker and Foster (2002)	"an entity with some form of claim on the focal organisation and sufficient power to influence that organisation"
Johnson and Scholes (2002)	"those individuals or groups who depend on the organisation to fulfil their own goals and on whom, in turn, the organisation depends"
Newcombe (2003)	"as groups or individuals who have a stake in or an expectation of the project's performance and indicated that this would include people inside the project, e.g. designers and contractors, and people outside the project e.g. users and the community"
Walker et al. (2008)	"an individual or group who has an interest or some aspect of rights or ownership in a project, and can contribute to, or be impacted by, either the work or the outcomes of the project"

identification of stakeholders appears in the definitions provided by Gupta (1995), Allen and Kilvington (2002), Mushove and Vogel (2005), Jepsen and Eskerod (2009), Reed (2008) and the WHO (2009). Stakeholder power is referred to in Gupta (1995), whilst the notion of relationships or interrelationships is covered in Gupta (1995), Brugha and Varvasovszky (2000) and Mushove and Vogel (2005). Schmeer (1999), Brugha and Varvasovszky (2000), Mushove and Vogel (2005) and Weible (2006) cover the importance of stakeholder interests, whilst the topic of influence is raised in Mushove and Vogel (2005) and Weible (2006). Only once is stakeholder analysis referred to as a management tool.

The various definitions of 'stakeholder analysis' can be seen below, in Table 6.4.

Growing importance

Stakeholder analysis has been described as having a "growing popularity" (Brugha and Varvasovszky, 2000) and as being critical to programme success (Nutt, 2002; Project Management Institute, 2004; Smudde and Courtright, 2011). It is argued that stakeholder analyses are "arguably more important than ever because of the increasingly interconnected nature of the world" (Bryson, 2004), that the importance

Table 6.4 Definitions of the term 'stakeholder analysis' within the literature

Author	*Definition of 'stakeholder analysis'*
Gupta (1995)	"to **identify** and **specify** the stakeholders and their interests, domain and specificity; identify and describe the **power relationships** between the stakeholders and the firm, and among the stakeholders; incorporate the concepts of **action and time**"
Schmeer (1999)	"a process of systematically gathering and analysing qualitative information to determine whose **interests** should be taken into account when developing and/or implementing a policy or programme"
Brugha and Varvasovszky (2000)	"an approach, a tool or set of tools for generating **knowledge about actors** so as to understand their behaviour, intentions, **interrelations and interests**; and for assessing the **influence and resources** they bring to bear on decision-making or implementation processes"
Allen and Kilvington (2002)	"the **identification** of a project's key stakeholders, an **assessment** of their interests, and the ways in which those interests affect project **riskiness and viability**"
Mushove and Vogel (2005)	"a range of tools or an approach for understanding a system by **identifying the key actors** or stakeholders on the basis of their attributes, **interrelationships** and assessing their respective **interests** related to the system, issue or resource"
Weible (2006)	"to address a set of questions: **who** are the stakeholders to include in the analysis; what are the stakeholders' **interests and beliefs**; who **controls** critical resources; with whom do stakeholders form **coalitions**; and what **strategies** and venues do stakeholders use to achieve their objectives"
Jepsen and Eskerod (2009)	"**identification** of stakeholders; **characterisation** of the stakeholders; decision about which strategy to use to **influence** each stakeholder"
Reed (2008)	"a process that: defines aspects of a social and natural system, **identifies** stakeholders, and **prioritises** stakeholders for involvement in the decision-making process"
World Health Organisation (2009)	"to **identify** stakeholders that will **influence** your project; anticipate the kind of influence, positive or negative, these groups will have on your project; develop **strategies** to get the most effective support possible for your project and **reduce any obstacles** to successful implementation"
Shirey (2012)	"Stakeholder analysis is an organisational **management tool** used for long-range strategic **planning** and stakeholder management. Stakeholder analysis can be used to inform project **planning, implementation, and evaluation**"

of stakeholder analysis is growing due to the governmental and non-profit reforms across the world (Barzelay, 2001; Kettl, 2002) and that "failure to attend to the information and concerns of stakeholders clearly is a kind of flaw in thinking or action that too often and too predictably leads to poor performance, outright failure, or even disaster" (Bryson, 2004). Jones (2003) adds that it is of importance in urban development projects to "stress exactly who the stakeholders are", whilst Lawson and Kearns (2010) add that it is only with the capturing of stakeholders' real interests, that they can be sufficiently empowered in urban development decision making.

Carrying out a stakeholder analysis

Various methods and techniques outlining how to carry out a stakeholder analysis have been outlined previously within the literature.

These typically follow a set amount of "steps" to follow. Such methods include Schmeer's "eight steps" (Schmeer, 1999), the five steps of Grimble and Chan (1995), the four steps of (1) stakeholder identification, (2) mapping of stakeholder interests, (3) evaluation of stakeholder importance and influence and (4) development of stakeholder strategy plan (Rietbergen-McCracken and Narayan, 1998; WHO, 2009), Shirey's (2012) four steps, the three steps of Reed et al. (2009) or Jepsen and Eskerod (2009) or the two steps of Jones (2003).

In addition, such models exist in aiding stakeholder analysis such as the:

- Influence-interest grid (Imperial College London, 2007)
- Power versus interest/predictability grids (Eden and Ackermann, 1998)
- Power-impact grid (Office of Government Commerce, UK, 2003)
- Three techniques used by the Australian Department of Sustainable Environment (2007) which include an influence-importance grid, Venn diagrams and CLIP (collaboration/conflict, legitimacy, influence and power) analysis
- Power-interest grid (Moorhouse Consulting, 2007)
- Problem-frame stakeholder mapping technique (Anderson et al., 1999)
- The three-dimensional grouping of power, interest and attitude (Murray-Webster and Simon, 2006)

Table 6.5, taken directly from Yang (2014), also shows practical methods for stakeholder analysis used in previous studies.

Towards a new method

As anticipated, the focus of this book is that of offering preliminary insights in a novel methodology, that of SuHousingImpact, by furthering an existing assessment method, SuROI. This will be achieved by creating a 'hybrid' method, aimed at drawing from the stakeholder analysis to sharpen the understanding of measurable benefits and costs for each stakeholder. The benefit of merging two different approaches will come from the filling of one another's gaps, due to the intrinsic nature of either qualitative or quantitative methods.

From one side, the stakeholder analysis presents some weaknesses: (1) Subjectivity. It has been noted that subjectivity can weaken the effectiveness of a stakeholder analysis (Shirey, 2012). The quantitative approach outlined within SuROI could be utilised to reduce the subjectivity of the stakeholder analysis. (2) Lack of outcomes. Bryson (2004) has argued that at present there is no overwhelming

Table 6.5 Practical methods for stakeholder analysis (Yang, 2014)

Approaches	*Description*	*Steps*		*Scholars*
		Stakeholder identification	*Stakeholder prioritisation*	
Focus groups	A small group brainstorm stakeholders, their interests, influence and other attributes and categorise them	✓	✓	Larson et al., 2010; Lawson and Kearns, 2010; Reed et al., 2009
Interviews	Interviews with stakeholders to identify their interests	✓		Mushove and Vogel, 2005; Oakley, 2007
Power/interest matrix (power/predictability matrix; stakeholder intensity index; the stakeholder impact index)	Despite various transformations, basically, stakeholders are categorised according to the levels of their power and interest		✓	Chiniyio and Akintoye, 2008; De Lopez, 2001; Orlander and Landin, 2008; Reed et al., 2009; Walker et al., 2008; Winch and Bonke, 2002
Snow-ball sampling	Based on identified stakeholders, a series of interviews and questionnaire surveys are conducted to identify more stakeholders	✓		Prell et al., 2009; Reed et al., 2009
Social network analysis	Through the use of structured interview/ questionnaire surveys, stakeholders' interrelationships are mapped and stakeholders' influence are analysed	✓	✓	Prell et al., 2009; Rowley, 1997
Stakeholder circle methodology	An integrated process for classifying stakeholders, prioritising stakeholders, visualising stakeholders and developing strategies	✓	✓	Bourne, 2005; Walker et al., 2008
Surveys	Relatively large numbers of stakeholders are expressed to express their opinion	✓		Prell et al., 2009; Timur and Getz, 2008
Workshops	Stakeholder representatives discuss specific issues and provide feedback	✓		Amado et al., 2009; Department of Planning, 2005

body of evidence indicating that stakeholder analyses help to produce desirable outcomes. Therefore, an addition is needed to the stakeholder analysis process showing a potential quantified outcome, thus proving that such an analysis can produce a desirable outcome. (3) Consider levels of interest/influence. Reed et al. (2009) proposed the use of 'extendable matrices' that consider levels of interest and influence of stakeholders. This book suggests that not only should the interest and influence of stakeholders be quantified, but how much a particular stakeholder would stand to gain, or lose from a given project, should also be quantified.

From the opposing side, SuROI still presents room for improvement: (1) Understanding of the "extent of stakeholders". As stated in Higham et al. (2017), Watson and Whitley (2016) suggest that with SROI, and so consequently with SuROI, being very stakeholder orientated, without an understanding of the extent of the "web of interconnected stakeholders", it cannot be clear at the outset if the analysis can capture the full nature of a project's eventual impact. This "understanding" can potentially be improved by reinforcing the stakeholder analysis component of the SuROI methodology. (2) Understanding who the winners and losers are. SuROI doesn't take into account who the winners and losers of a given project are, and how much they gain or lose from a project individually – it normally allows quantifying only the overall benefits from a given intervention.

The proposed 'SuHousingImpact' method is transdisciplinary in nature, in that it combines the quantitative approach of SuROI (which stems from cost-benefit analysis) with the qualitative approach of stakeholder analysis (which in essence is a management/policy making tool). The creation of a 'hybrid' method allows the production of a new 'method', useful to support decision-makers not only in making the right decisions, but also in negotiating with potential interested parties. Being a management/policy making tool, current stakeholder analyses fall into the same category as previous methods of evaluating housing-led urban regeneration projects, which is that they are typically qualitative in nature. As has been seen earlier in this book, qualitative methods have disadvantages, which can be solved by introducing quantities into the analysis. To counteract the qualitative essence of stakeholder analysis, we can use the SuROI methodology to change the analysis into a quantifiable one. By doing this, it would be possible to quantifiably measure how much each given stakeholder of a project gains or losses from that given project. At the same time, by readdressing the SuROI traditional methodology towards a stakeholder-oriented perspective, it is possible to emphasise the potential benefits for different stakeholders influenced by the intervention, thus orienting its usability from a decision making tool into a management tool. This is particularly relevant in the era of 'New Public Management' within public administration which involves such aspects as explicit standards and measures of performance, a greater emphasis on output controls, a shift to the disaggregating of units in the public sector, a shift to greater competition in the public sector and a stress on greater discipline in resource use (Hood, 1991). New Public Management additionally involves many different administrative departments, each with their own separate budgets, competing against each other for potential financial injections, in a similar way to the "new territorial governance" described by Roberts (2000). Indeed, Roberts (2006) asserts that the allocation of power, responsibilities and resources is more fragmented than in the past.

SuROI assumes that there is one large public domain, whereas in practice this is not the case. New Public Management correctly assumes that there are competing agencies where the behaviour of the differing agencies' management is changing. Currently it is the case through New Public Management that each separate stakeholder has its own budget and each separate stakeholder, such as the National Health Service (NHS), highways or parks and gardens to name but only a few mere examples, is fighting for as much money as it itself can get hold of, and probably looks at its own needs before thinking about needs belonging to other competing stakeholders.

Therefore, rather than thinking in terms of overall 'public' benefit, it is important to rely on quantifiable methods showing benefits and costs for different public stakeholders. This would potentially be able to show how a specific intervention reflects on each separate stakeholder's budget line.

The prospective 'SuHousingImpact' tool would be used to gauge the quantifiable amount of impact for each stakeholder involved in a housing-led urban regeneration scheme and would show in monetary units whether or not each stakeholder gains or losses from a given project, and additionally, by the cash flow amount involved.

This aspect of the new method could then potentially act as a negotiating tool in the quest for more investment into housing-led urban regeneration schemes, thus ensuring financial sustainability going forward.

Figure 6.1 acts as a pictorial example of what is meant; with impact shown to include the police and NHS (the publicly funded healthcare system in the United Kingdom which was formed in 1948).

Concept of time

In addition, there is further potential of the SuROI approach as a quantitative methodology for the assessment of interventions. In particular, the concept of time can be brought into SuROI. As Roberts (2006) asserts, "it is unlikely that all elements of a sustainable development programme will be delivered at a single point in time".

Figure 6.1 Quantification of "winners and losers" using the SuROI methodology (author's elaboration).

The payback period is one of the most popular methods in the evaluation of capital budgeting decisions which has been referred to as "one of the most important techniques in evaluating capital budgeting" (Mawih, 2015) and can be brought into play as part of the novel attributes of the method. It is a simple method that determines when a project breaks even (Awomelwe and Ogundele, 2008) and has been defined as the time required to recover the initial investment in a project from operations (Ibid, 2008).

The payback period has been defined as the number of years it would take to recover a project's costs of investment and is in use all over the world, including in the United Kingdom and the USA where the payback period has been shown to be an important method (Pike, 1985), in Japan (Shinoda, 2010), in Kenya (Khakasa, 2009) and indeed on various different continents of the globe, including Europe, America and Africa (Awomelwe and Ogundele, 2008). Managers have been said to be "risk averse" and so typically attempt to minimise such risk by using the payback period method (Mawih, 2015).

The payback period can be easily expressed through the following equation:

$$\text{Payback period} = \frac{\text{cost of project}}{\text{annual cash flow}}$$

The popularity of the payback period has been cited by Ryan and Ryan (2002), who comment, as part of their research looking into the uses of capital budgeting techniques by 1000 chief financial officers and financial managers, that one of their principal conclusions is that financial managers prefer to use multiple capital budgeting techniques including that of the payback period. As Blatt (1979) had earlier remarked, "a payback time limit is used by nearly every responsible manager, in fact". Segelod (1995) cites that although the use of payback period as a single criterion has decreased over time, its use as a secondary measure has actually increased over the same period.

Some advantages and disadvantages of the payback period include:

Advantages:

- The payback period technique is simple and easy to understand (Dayananda et al., 2002; Mawih, 2015), being probably the simplest method to use to rank projects in accordance of timestreams of costs and benefits (French, 1988).
- There is a wide acceptance of this technique by managers and investors (Mawih, 2015). Indeed, data from numerous previous studies show that managers "prefer the simple payback period method" (Shinoda, 2010).
- It has a great importance, which includes, but is not limited to, its simplicity, liquidity and risk assessment (Awomelwe and Ogundele, 2008).
- It is still used in organisations all over the world (Ibid, 2008).
- It contains a built-in safeguard against risk and uncertainty in that the earlier the payback the lower the risk (Ibid, 2008).
- It remains a major supplementary tool in investment analysis (Ibid, 2008).

Disadvantages:

- It ignores cash flows (Mawih, 2015).

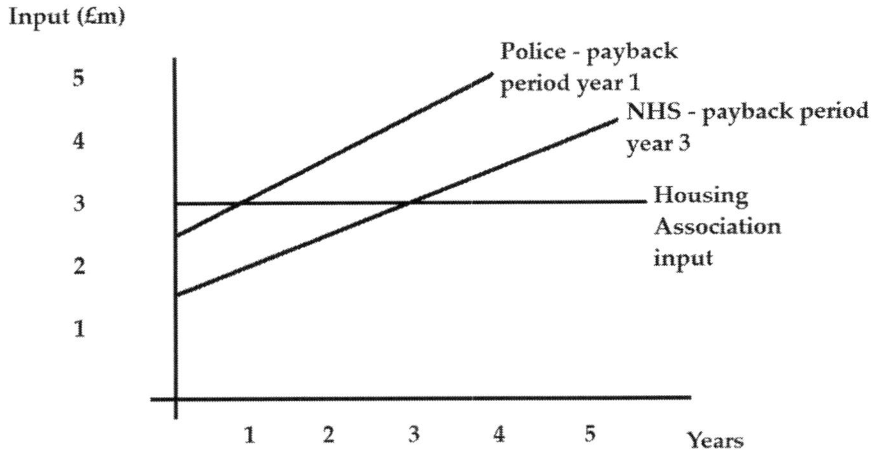

Figure 6.2 Payback period by using calculations from the SuROI methodology (author's elaboration).

- It cannot distinguish between projects with the same payback period (French, 1988).
- It ignores any benefits that occur after the payback period, i.e. it does not measure total income (Awomelwe and Ogundele, 2008).

The introduction of the notion of the payback period being available for each individual stakeholder can enable the provision of a rough estimate of when each involved stakeholder could expect to see a gain from a housing-led intervention (Figure 6.2).

Summary

This chapter has explained exactly how a new stakeholder-led evaluation method – the 'SuHousingImpact' tool which has been developed – functions. Gaps within SuROI are explained, whilst the MSP approach, Multi-Stakeholder Networks, Stakeholder Management and the subject area of Stakeholder Analysis, all areas which link in to the new evaluation tool, were explained, together with the theoretical underpinning of the method and how the tool relates directly to common budgetary issues found within the concept of New Public Management.

The next chapter explains the SuHousingImpact method stage by stage, with explanatory guidance so academics and urban regeneration professionals are able to actively and regularly use it.

7 The SuHousingImpact method explained

Introduction

This chapter will explain exactly how the method (the 'SuHousingImpact' tool), which has been developed, functions. Screenshots of the method together with step-by-step instructions will be provided in order that readers of this book, including academics, students, public sector employees and housing association staff alike, are able to fully understand and subsequently implement the method for their own use. Guidance will also be given on sources of indicators and proxies to feed the tool with all necessary quantifiable information on socio-environmental spillovers from housing-led urban regeneration schemes. The tool itself can be accessed via www.suhousingimpact.com.

The SuHousingImpact method

It is prudent at this stage to be reacquainted with the method itself (Figure 7.1).

As can be seen, and as has been stated previously, the method is a refinement of SuROI methodology.

The method actively uses not only the SROI (Social Value UK) impact map, in spreadsheet format, but additionally uses the principles of SROI.

In addition, the method uses SuROI quantitative measurement of social and environmental impacts and spillovers, and potential usage of SROI, ESA (Ecosystem Services Analysis), Wellbeing Valuation and/or any other method of gauging social and environmental impact quantitatively, that will fit flexibly into the impact map spreadsheet.

The 'SuHousingImpact' tool then subsequently refines and furthers SuROI methodology by creating a novel 'stage 6' to the impact map process.

By way of this new 'stage 6', the 'SuHousingImpact' tool then focuses on the potential to look in far greater depth at future economic sustainability, than has previously been covered within SuROI. By doing this, the three separate aspects of the triple bottom line are all being actively looked at within this method and future economic sustainability is being covered more actively than has been previously carried out by way of SuROI.

DOI: 10.1201/9781003043881-7

Stage 1 →		Stage 2 →			
Stakeholders	**Intended/unintended changes**	**Inputs**		**Outputs**	**The Outcomes (What changes)**
					Description
Who do we have an effect on? Who has an effect on us?	What do you think will change for them?	What do they invest?	Value £	Summary of activity in numbers	How would you describe the change?

Stage 3 →							Stage 4 →					
							Dead-weight %	**Displace-ment %**	**Attribu-tion %**	**Drop Off %**	**Impact**	**Action Plan**
Indicator	Source	Quantity	Duration	Financial proxy	Value £	Source	What would have happened without the activity?	What activity did you displace?	Who else contributed to the change?	Does the outcome drop off in future years?	Quantity times financial proxy, less deadweight, displacement and attribution.	What needs to be done
How would you measure it?	Where did you get the information from?	How much change was there?	How long does it last?	What proxy would you use to value the change?	What is the value of the change?	Where did you get the information from?						

		Stage 5 →					
		Calculating Social Return					
		Discount rate		**3.5%**			
		Year 0	**Year 1**	**Year 2**	**Year 3**	**Year 4**	**Year 5**
		0.00	200.00	0.00	0.00	0.00	0.00
		0.00	0.00	0.00	0.00	0.00	0.00
		0.00	400.00	0.00	0.00	0.00	0.00
		0.00	0.00	0.00	0.00	0.00	0.00
		0.00	0.00	0.00	0.00	0.00	0.00
		0.00	0.00	0.00	0.00	0.00	0.00
		0.00	0.00	0.00	0.00	0.00	0.00
		0.00	0.00	0.00	0.00	0.00	0.00
		0.00	0.00	0.00	0.00	0.00	0.00
		0.00	0.00	0.00	0.00	0.00	0.00
		0.00	0.00	0.00	0.00	0.00	0.00
Total	600.00	0.00	600.00	0.00	0.00	0.00	0.00
Present value of each year		0.00	579.71	0.00	0.00	0.00	0.00
Total Present Value (PV)							579.71
Net Present Value							379.71
(PV minus the investment)							
Social Return							2.90
Value per amount invested							

Figure 7.1 The 'SuHousingImpact' method.

Stakeholder E	Cash inflow	Cash outflow	Cumulative cash flow
Y0	0	−100	−100
Y1	£ 152.05	0	£ 52.05
Y2	£ 146.73	0	£ 198.78
Y3	£ 141.59	0	£ 340.37
Y4	£ 136.64	0	£ 477.01
Y5	£ 131.85	0	£ 608.86

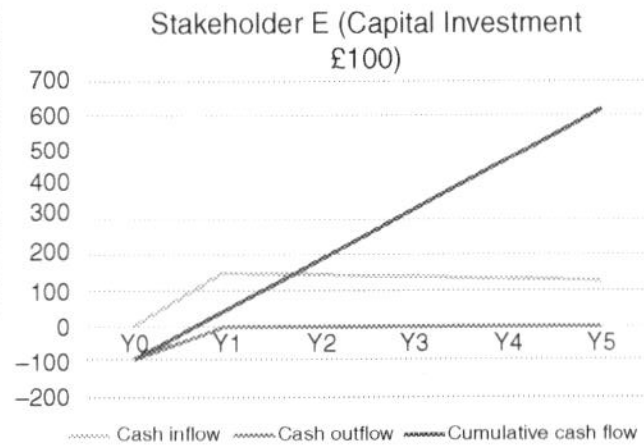

Stakeholder F	Cash inflow	Cash outflow	Cumulative cash flow
Y0	0	−25000	−25000
Y1	£ 966,768.00	0	£ 941,768.00
Y2	£ 932,931.12	0	£ 1,874,699.12
Y3	£ 900,278.53	0	£ 2,774,977.65
Y4	£ 868,768.78	0	£ 3,643,746.43
Y5	£ 838,361.87	0	£ 4,482,108.31

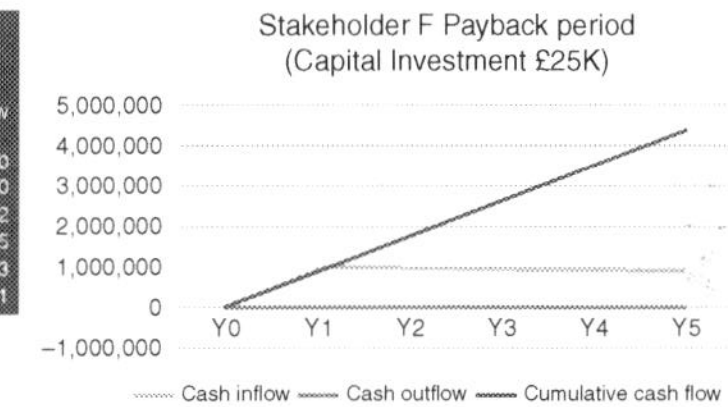

Stakeholder E	Cash inflow	Cash outflow	Cumulative cash flow
Y0	0	−500	−500
Y1	£ 152.05	0	-£ 347.95
Y2	£ 146.73	0	-£ 201.22
Y3	£ 141.59	0	-£ 59.63
Y4	£ 136.64	0	£ 77.01
Y5	£ 131.85	0	£ 208.86

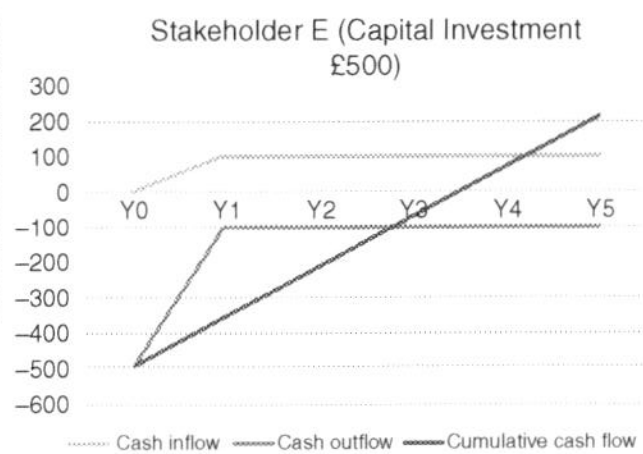

Stakeholder F	Cash inflow	Cash outflow	Cumulative cash flow
Y0	0	−250000	−250000
Y1	£ 966,768.00	0	£ 716,768.00
Y2	£ 932,931.12	0	£ 1,649,699.12
Y3	£ 900,278.53	0	£ 2,549,977.65
Y4	£ 868,768.78	0	£ 3,418,746.43
Y5	£ 838,361.87	0	£ 4,257,108.31

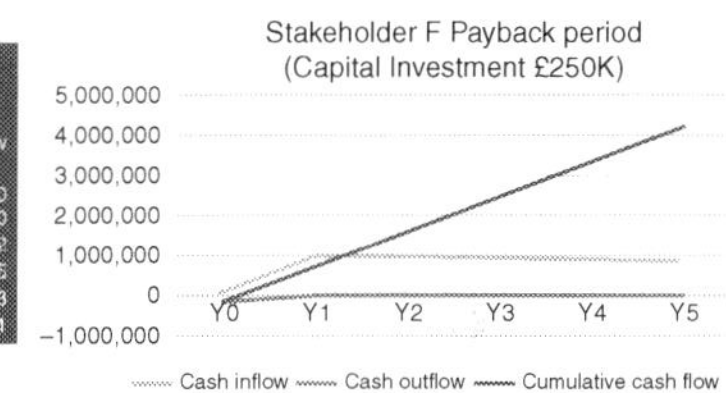

Figure 7.1 (Continued)

Whole scheme	Cash inflow	Cash outflow	Cumulative cash flow
Y0	0	−5000000	−5000000
Y1	£ 5,332,685.29	0	£ 332,685.29
Y2	£ 5,146,041.30	0	£ 5,478,726.59
Y3	£ 4,965,929.86	0	£ 10,444,656.45
Y4	£ 4,792,122.31	0	£ 15,236,778.77
Y5	£ 4,624,398.03	0	£ 19,861,176.80

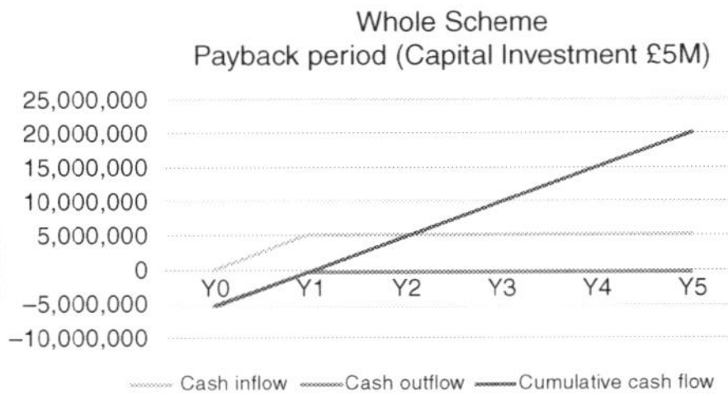

Whole scheme	Cash inflow	Cash outflow	Cumulative cash flow
Y0	0	−10000000	−10000000
Y1	£ 5,332,685.29	0	£ 4,667,314.71
Y2	£ 5,146,041.30	0	£ 478,726.59
Y3	£ 4,965,929.86	0	£ 5,444,656.45
Y4	£ 4,792,122.31	0	£ 10,236,778.77
Y5	£ 4,624,398.03	0	£ 14,861,176.80

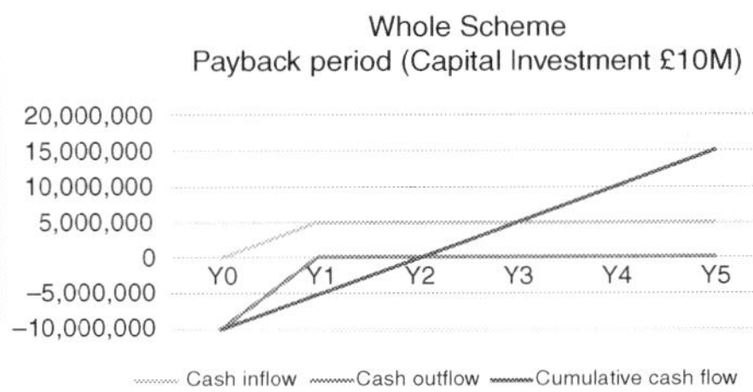

Figure 7.1 (Continued)

This active economic focus is carried out through the provision of a platform within which costs and benefits to potential future stakeholders can be mapped out; such stakeholders being potentially involved in the housing-led urban regeneration scheme or process. The SuROI assessment methodology has been easily adapted by being refocused in terms of stakeholders' engagement. By repacking the financial calculation allocating costs and benefits across the different stakeholders involved, it would be possible to attract potential new investors, willing to increase the benefits that the method has unveiled.

If previously invisible and intangible benefits resulting from a housing-led urban regeneration scheme are to be made visible and tangible by using the SuROI method, then by analysing which potential stakeholders are receiving benefit from the said scheme, we can highlight to those stakeholders not yet involved in the scheme, the cost benefit that is attributable and which is actively helping that stakeholder organisation, as a result of the said scheme.

By actively seeking out and discussing the financial benefits that are coming the way of the not yet involved stakeholder, it is possible to prevent schemes being affected by current economic cuts and subsequently ensuring that there are adequate resources to match the ambition of sustainable development (Roberts, 2000) and the United Nations (UN) SDGs (Sustainable Development Goals), in particular Goal 11 of the SDGs, by offsetting budgetary cuts through striking agreements with those stakeholders who benefit, thus tapping into the theoretical premise covered within the Kaldor-Hicks criterion.

As an example, let us say that due to a housing-led urban regeneration scheme, local housing association tenants are benefitting health-wise from works that have been carried out, but the Healthcare Provider Y is not currently an active stakeholder in the scheme in question. Through using the 'SuHousingImpact' tool, it can be highlighted to the Healthcare Provider Y on how much they are potentially receiving in terms of previously intangible benefits being costed out, according to the methodology, which results in savings to the Government.

It can also be highlighted to Healthcare Provider Y, that although they do not have to participate in any way in the urban regeneration scheme in question, by contributing an amount of money to the ongoing works involved in the said scheme, they could be offsetting the financial contribution by saving an outlay of money in the future. It could be the case that by not investing, they accrue higher costs if the sustainability of the said scheme is compromised by a lack of funds, which then ends the productivity of the scheme and ends the benefits that were previously being created; those benefits previously saving the Healthcare Provider Y money.

Research previously carried out with practitioners in the housing field was the basis of this new method. Previously, a great deal of investment has been put into many regeneration schemes, only for social (and environmental) return to be minimal. It has also been relayed that in the current economic climate, housing associations do not want to be outlaying massive investment, only to be having to do exactly the same in 30 years' time. It has been found that more sustainability, less wasting of money and, most importantly, ways to plug the gap created by the recently experienced economic shortfall which is impacting on housing associations, tenants and communities alike are important and needed (Dean, 2019).

Focusing practically on the SuHousingImpact tool, calculations and formulas have been inputted into the impact map spreadsheet to split potential gains and losses in pounds sterling between individual and different stakeholders, aligning each separate impact with each individual stakeholder, whilst also, additional formulas have then been inputted and used within the spreadsheet, to calculate payback period analyses for each individual potential stakeholder. The thinking behind this being that if one or more potential stakeholders are being found to be profiting well from a given scheme, then the monetary values within the spreadsheet can act as evidence to be supplied to the stakeholder(s) in question, which may then enable a cost contribution via agreement to be made by the stakeholder(s) in question to the housing association or housing-led urban regeneration facilitator in question. This can then potentially mitigate financial problems or concerns on behalf of the housing association or facilitator in question during a period of current and severe economic constraints.

Lastly, in the same way as payback periods for each individual stakeholder have been carried out, the same payback period calculations can also be performed for the scheme as a whole.

In terms of how this has been technically carried out within the impact map spreadsheet, firstly the involved formulas which enable these sums to be processed will be discussed, before moving on to commentary and guidance on the six stages involved in the tool.

1 Socio-environmental impact per stakeholder

If we use the example below, which is presented within an impact map with only two hypothetical stakeholders of Housing Association X customers' and Healthcare Provider Y, for ease of viewing, clarity and to aid learning, it can be seen that formulas can be inputted within the impact map spreadsheet in order to separate socio-environmental spillovers that previously although being a part of the SuROI spreadsheet calculations, were not organised or highlighted in this specific way, that is to say per individual stakeholder.

In the above screenshot, it can be observed that an additional 'stage 6' has been added to the existing SuROI impact map. Formulas have then been inputted to provide monetary valuations of socio-environmental spillovers on a per stakeholder basis, which then over time (years one to five) take into account the 3.5% depreciation rate already advised by HM Treasury's Green Book, and utilised previously within both SROI and SuROI.

The screenshot in Figure 7.2 shows this 3.5%, which is added to the initial impact calculation, year on year, meaning that the value of impacts drop off over years one to five. The total amount of socio-environmental spillovers (the cash inflow column within the table) comprises of the addition of all spillovers attributed to the particular stakeholder in question – on this occasion, the stakeholder of Housing Association X.

Housing Association X	Cash inflow	Cash outflow	Cumulative cash flow
Y0	0	−100000	−100000
Y1	£ 912,320.24	0	£ 812,320.24
Y2	£ 880,389.03	0	£ 1,692,709.27
Y3	£ 849,575.42	0	£ 2,542,284.69
Y4	£ 819,840.28	0	£ 3,362,124.96
Y5	£ 791,145.87	0	£ 4,153,270.83

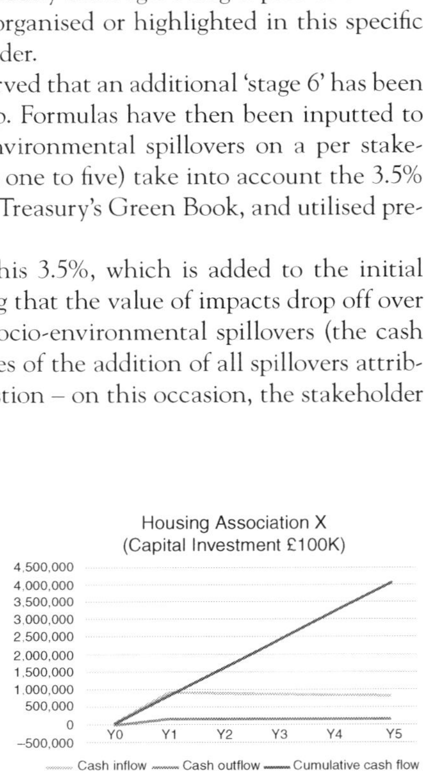

Housing Association X	Cash inflow	Cash outflow	Cumulative cash flow
Y0	0	−500000	−500000
Y1	£ 912,320.24	0	£ 412,320.24
Y2	£ 880,389.03	0	£ 1,292,709.27
Y3	£ 849,575.42	0	£ 2,142,284.69
Y4	£ 819,840.28	0	£ 2,962,124.96
Y5	£ 791,145.87	0	£ 3,753,270.83

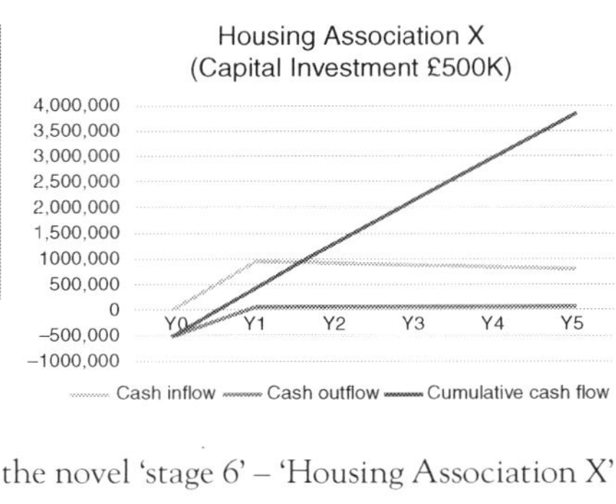

Figure 7.2 The 'SuHousingImpact' method – the novel 'stage 6' – 'Housing Association X' stakeholder calculations.

This principle can again then be utilised for the next involved stakeholder – that of Healthcare Provider Y. This method can then subsequently be repeated for the many stakeholders often involved in housing-led urban regeneration schemes (Figure 7.3).

At the end of such a process, a comprehensive list can be built of all involved stakeholders and their respective socio-environmental impacts from a given housing-led urban regeneration scheme.

The way the discounted figure is built up over time can be seen from the formulas below. The same yearly depreciation percentage is added year on year to provide a discounted impact value per year, from years one to five (Figures 7.4–7.7):

The same formula can be used with all additional stakeholders.

Healthcare Provider Y	Cash inflow	Cash outflow	Cumulative cash flow
Y0	0	−25000	−25000
Y1	£ 966,768.00	0	£ 941,768.00
Y2	£ 932,931.12	0	£ 1,874,699.12
Y3	£ 900,278.53	0	£ 2,774,977.65
Y4	£ 868,768.78	0	£ 3,643,746.43
Y5	£ 838,361.87	0	£ 4,482,108.31

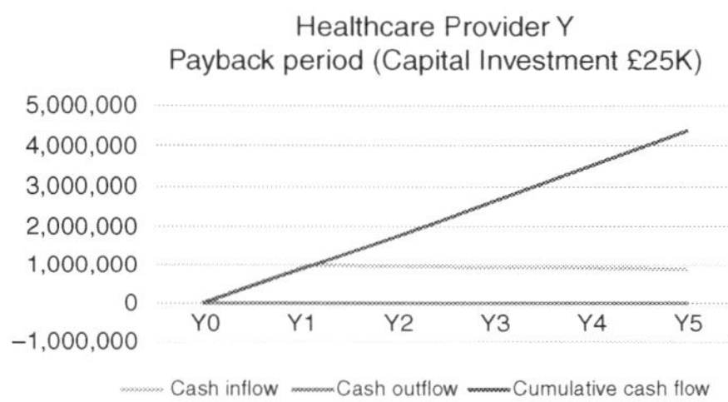

Healthcare Provider Y	Cash inflow	Cash outflow	Cumulative cash flow
Y0	0	−250000	−250000
Y1	£ 966,768.00	0	£ 716,768.00
Y2	£ 932,931.12	0	£ 1,649,699.12
Y3	£ 900,278.53	0	£ 2,549,977.65
Y4	£ 868,768.78	0	£ 3,418,746.43
Y5	£ 838,361.87	0	£ 4,257,108.31

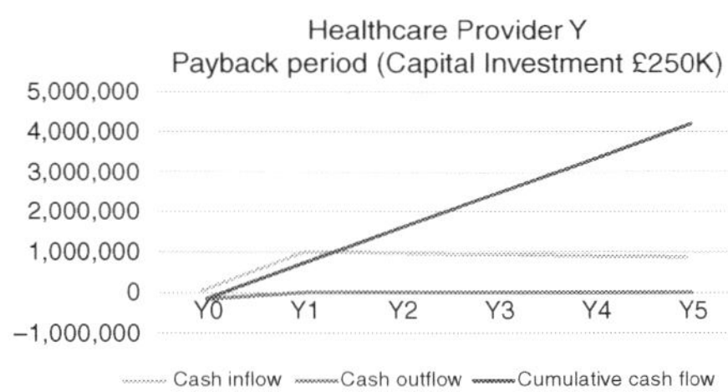

Figure 7.3 The 'SuHousingImpact' Method – the novel 'stage 6' – 'Healthcare Provider Y' stakeholder calculations.

NHS	Cash inflow	Cash outflow	Cumulative cash flow
Y0	0	−25000	−25000
Y1	£ 966,768.00	0	£ 941,768.00
Y2	£ 932,931.12	0	£ 1,874,699.12
Y3	£ 900,278.53	0	£ 2,774,977.65
Y4	£ 868,768.78	0	£ 3,643,746.43
Y5	£ 838,361.87	0	£ 4,482,108.31

Figure 7.4 Year 1 of depreciation.

NHS	Cash inflow	Cash outflow	Cumulative cash flow
Y0	0	−25000	−25000
Y1	£ 966,768.00	0	£ 941,768.00
Y2	£ 932,931.12	0	£ 1,874,699.12
Y3	£ 900,278.53	0	£ 2,774,977.65
Y4	£ 868,768.78	0	£ 3,643,746.43
Y5	£ 838,361.87	0	£ 4,482,108.31

Figure 7.5 Year 2 of depreciation.

NHS	Cash inflow	Cash outflow	Cumulative cash flow
Y0	0	−25000	−25000
Y1	£ 966,768.00	0	£ 941,768.00
Y2	£ 932,931.12	0	£ 1,874,699.12
Y3	£ 900,278.53	0	£ 2,774,977.65
Y4	£ 868,768.78	0	£ 3,643,746.43
Y5	£ 838,361.87	0	£ 4,482,108.31

Figure 7.6 Year 3 of depreciation.

NHS	Cash inflow	Cash outflow	Cumulative cash flow
Y0	0	−25000	−25000
Y1	£ 966,768.00	0	£ 941,768.00
Y2	£ 932,931.12	0	£ 1,874,699.12
Y3	£ 900,278.53	0	£ 2,774,977.65
Y4	£ 868,768.78	0	£ 3,643,746.43
Y5	£ 838,361.87	0	£ 4,482,108.31

Figure 7.7 Year 4 of depreciation.

2 Payback period analyses:

The payback period has been defined as the number of years it would take to recover a project's costs of investment. Simple payback period analyses can be carried out using the method by comparing the impact as against the input value per year either for each individual stakeholder or within a scheme as an entire entity.

The socio-environmental spillover values created from the earlier calculations can be used in comparison with various levels of capital investment values. The subsequent results can be provided in tabular and graphical format within 'stage 6' of the method.

Different capital investment amounts can be inputted into the tables (such as £25,000 and £250,000 as per the example in Figure 7.8), which then promotes different payback period analyses, on a per stakeholder basis, according to the amount of money a particular stakeholder is willing to contribute to a scheme. The financial amount potentially inputted by a particular stakeholder can be offset by the monetary value received by the said stakeholder in the form of the cash inflow, which is the amount of monetisable socio-environmental spillovers.

In addition, the same calculations can be used for the scheme as a whole, as in Figure 7.9, taken from the Housing Association X environmental-led programme:

Healthcare Provider Y	Cash inflow	Cash outflow	Cumulative cash flow
Y0	0	−25000	−25000
Y1	£ 966,768.00	0	£ 941,768.00
Y2	£ 932,931.12	0	£ 1,874,699.12
Y3	£ 900,278.53	0	£ 2,774,977.65
Y4	£ 868,768.78	0	£ 3,643,746.43
Y5	£ 838,361.87	0	£ 4,482,108.31

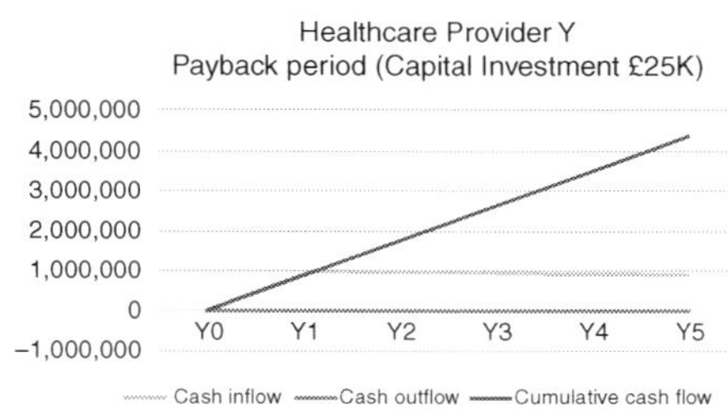

Healthcare Provider Y	Cash inflow	Cash outflow	Cumulative cash flow
Y0	0	−250000	−250000
Y1	£ 966,768.00	0	£ 716,768.00
Y2	£ 932,931.12	0	£ 1,649,699.12
Y3	£ 900,278.53	0	£ 2,549,977.65
Y4	£ 868,768.78	0	£ 3,418,746.43
Y5	£ 838,361.87	0	£ 4,257,108.31

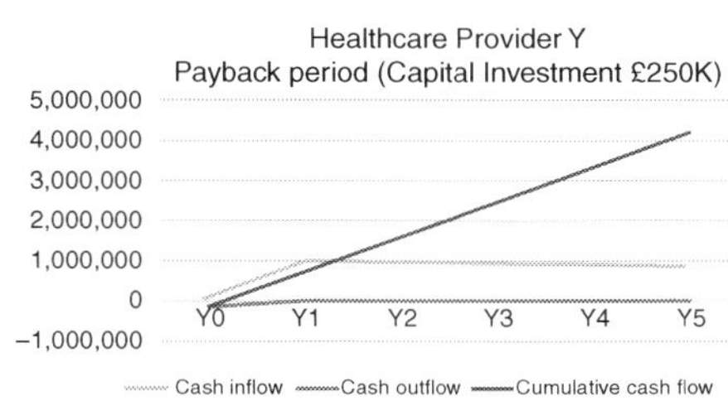

Figure 7.8 Spillovers versus different levels of capital investment for the 'Healthcare Provider Y' stakeholder.

Whole scheme	Cash inflow	Cash outflow	Cumulative cash flow
Y0	0	−5000000	−5000000
Y1	£ 5,332,685.29	0	£ 332,685.29
Y2	£ 5,146,041.30	0	£ 5,478,726.59
Y3	£ 4,965,929.86	0	£ 10,444,656.45
Y4	£ 4,792,122.31	0	£ 15,236,778.77
Y5	£ 4,624,398.03	0	£ 19,861,176.80

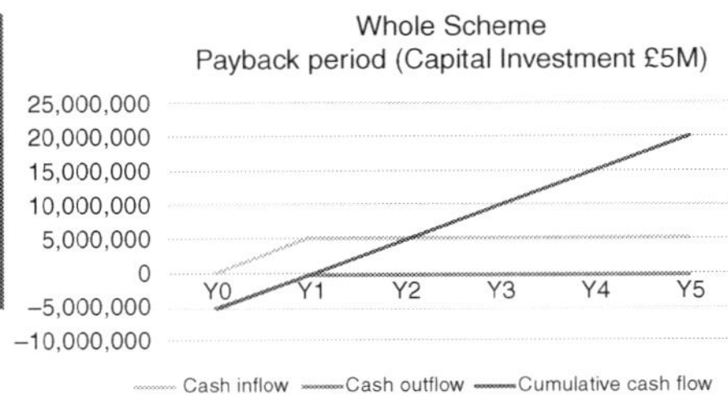

Whole scheme	Cash inflow	Cash outflow	Cumulative cash flow
Y0	0	−10000000	−10000000
Y1	£ 5,332,685.29	0	£ 4,667,314.71
Y2	£ 5,146,041.30	0	£ 478,726.59
Y3	£ 4,965,929.86	0	£ 5,444,656.45
Y4	£ 4,792,122.31	0	£ 10,236,778.77
Y5	£ 4,624,398.03	0	£ 14,861,176.80

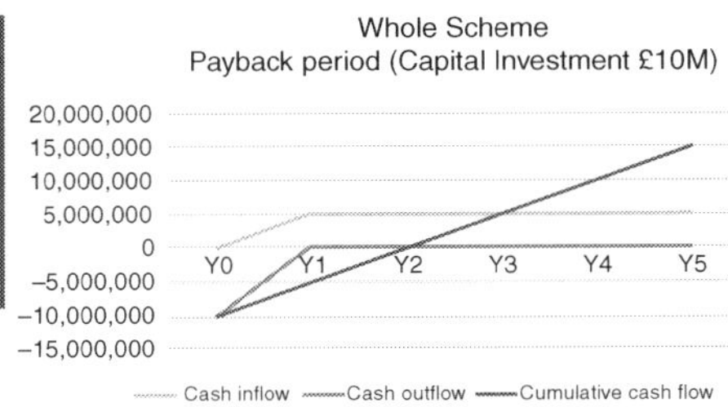

Figure 7.9 Payback period for environmental-led programme as a whole.

Again, the different amounts of money inputted (in this example £5m and £10m) can be used to gauge the various differing payback periods for each stakeholder (or in this example, for a scheme as a whole) and would be changeable according to the amount of cash offered.

How to use the tool – stage one to stage six

Firstly, before commencing on the use of the SuHousingImpact tool, it would be beneficial for any prospective user of the tool to familiarise themselves with the SROI or SuROI impact map and the accompanying SROI principles which underpin the valuation processes within such a methodology. These have been previously cited within this book, however for ease of reference, they are repeated below:

There are seven principles of SROI as follows (Nicholls et al., 2012):

- Involve stakeholders
- Understand what changes
- Value the things that matter
- Only include what is material
- Do not over-claim
- Be transparent
- Verify the result

In addition, the various stages of SROI can be seen below (adapted from Nicholls et al., 2012):

1 Establishing scope and identifying key stakeholders.
   Clarity is needed on exactly what the SROI will cover, who will be involved and how.
2 Mapping outcomes.
   Stakeholder engagement allows the development of the impact map and creates a theory of change which shows relationships between inputs, outputs and outcomes.
3 Evidencing outcomes and giving them a value.
   Researching of appropriate data to show whether outcomes have happened and subsequently valuing them.
4 Establishing impact.
   Once evidence is collected on outcomes and they are subsequently monetised, aspects of change that would have happened anyway are deducted.
5 Calculating the SROI.
   Calculations involve the addition of all benefits, before subtracting any negatives and comparing the result to the investment. This is also where the sensitivity of the results can be tested.

Stage one

The first stage of the SuHousingImpact tool, in the same way as with SROI and SuROI, involves the inputting of stakeholder information. Such information requires active thinking about which stakeholders are involved within a scheme. This initial stage is very important because without properly taking into account the correct number of stakeholders involved in a scheme, incorrect impact values will be produced and it will not be clear if the analysis can capture the full nature of a project's eventual impact (Higham et al., 2017; Watson and Whitley, 2016). This will have an impact on all figures from the overall socio-environmental impact, to the impacts per stakeholder and the payback period analyses. As previously advised within this book, this aspect can potentially be improved by reinforcing the stakeholder analysis component of the SuROI methodology. Once all stakeholders are listed, then all intended and unintended changes can be listed within the impact map spreadsheet. This is where one looks on change encountered for the stakeholders involved in the housing-led urban regeneration scheme in question. Written data should be filled in within the cells found under the stage one heading (Figure 7.10).

Stage two

Stage two of the process involves filling in information covering the inputs of a scheme, together with information on outputs and outcomes. Input information includes anything which is invested in a scheme, together with the value of those inputs, in currency. The outputs should involve numerical summaries,

Stage 1	
Stakeholders	**Intended/unintended changes**
Who do we have an affect on? Who has an effect on us?	What do you think will change for them?

Figure 7.10 Stage one of the SuHousingImpact method.

Stage 2			
Inputs		**Outputs**	**The Outcomes (What)**
What do they invest?	What is the value of the inputs in currency *(only enter numbers)*	Summary of activity in numbers	Description How would the stakeholder describe the changes?

Figure 7.11 Stage two of the SuHousingImpact method.

and outcome information should include how the involved stakeholders would describe the changes. It is important not to confuse the concepts of output and outcome. An example of an output would be an amount of jobs created or the amount of persons employed as a result of a given scheme, whilst related outcomes to these examples cited would include aspects such as personal well-being or confidence improving as a consequence of the aforementioned outputs (Figure 7.11).

Stage three

Stage three of the method involves assigning monetary values to the outputs and outcomes assigned in stage two. These monetary values are in the form of either indicators and/or proxy values. For each indicator, the source should be highlighted (information such as this satisfies the earlier mentioned transparency in SROI principle) together with the change and duration. In addition to indicator values, financial proxy values can then be inputted for valuation purposes together with the accompanying value in currency and the source of the proxy, again to promote transparency. A financial proxy can be described as a figure that can be assigned as a stand in for variables that cannot be directly measured. The value of change and quantity of change are calculated via the embedded formulas within the spreadsheet.

In terms of sourcing indicators and/or proxy values for inputting purposes, any related and good quality primary data can and should be used for these purposes. Ideally, direct communication with stakeholders should take place either in the form of interviews or focus groups. In terms of appropriate secondary data, the following sources are of use:

Housing Associations' Charitable Trust (HACT) Social Value Bank

The HACT Social Value Bank database measures how much people's life satisfaction changes and is based on large datasets from national UK surveys (Trotter et al., 2014) such as the British Household Panel Survey (BHPS), Understanding Society, The Crime Survey of England and Wales and The Taking Part Survey (Bichard, 2015). These surveys are carried out in a general manner and attempt to gauge how people feel about their lives overall. This general approach avoids any psychological biases involved with questions being posed about a particular scheme or a particular project (Ibid, 2015). Of late, the English Housing Survey has also been used to contribute to the database. This involves information from over 13,000 households, including interviews with householders and a physical survey being carried out on the resident's property in question.

The HACT database includes up to 74 proxy values and more are to be added in 2018 (Figure 7.12).

Code	Outcome	Unknown	<25	25-49	>50	London	<25	25-49	>50	Outside of London	<25	25-49	>50
EMP1401	• Full-time employment	£14,433	£13,446	£15,354	£13,865	£14,380	£10,126	£15,337	£16,168	£14,433	£13,702	£15,371	£13,720
EMP1602	• Self-employment	£11,588	£13,157	£11,952	£10,538	£12,116	£16,471	£12,406	£11,027	£11,537	£12,848	£11,887	£10,520
EMP1603	• Part-time employment	£1,229	£737	£1,824	£1,966	£1,966	£1,966	£1,966	£1,966	£1,176	£737	£1,850	£1,966
EMP1604	• Government training scheme	£9,447	£11,903	£5,668	£12,817	£5,668	£5,668	£5,668	£5,668	£9,596	£12,083	£5,668	£12,034
EMP1605	• Secure job	£12,034	£10,871	£12,148	£11,969	£10,502	£9,089	£11,504	£8,993	£12,083	£10,923	£12,164	£12,116
EMP1606	• Apprenticeship	£2,353	£1,861	£2,948	£3,090	£3,764	£3,764	£3,764	£3,764	£2,195	£1,756	£2,868	E2,985
EMP1607	• Vocational training	£1,124	£1,124	£1,124	£1,124	£1,798	£1,798	£1,798	£1,798	£1,019	£1,019	£1,019	E1,019
EMP1408	• Regular volunteering	£3,249	£2,895	£2,742	£3,324	£3,772	£4,663	£4,663	£2,306	£3,199	£2,562	£2,536	£3,474
EMP1609	• Regular attendance at voluntary or local organisation	£1,773	£1,901	£1,064	£1,064	£1,064	£2,836	£1,064	£1,064	£1,875	£1,824	£1,567	£1,824
EMP1610	• General training for job	£1,567	£2,507	£1,437	£2,507	£2,206	£940	£2,507	£2,507	£1,515	£2,507	£1,359	£2,507
EMP1611	• Employment training	£807	£484	£887	£940	£1,291	£1,291	£1,291	£484	£647	£484	£674	£1,071
EMP1612	• Employed parent for children (11-15)	£1,700	£1,700	#N/A	#N/A	£1,020	£1,020	#N/A	#N/A	£1,901	£1,901	#N/A	#N/A
ENV1601	» No problem with teenagers hanging around	£5,760	£6,963	£4,684	£8,746	£4,333	£3,456	£3,456	£9,216	£6,034	£7,532	£5,084	£8,353
ENV1602	» No problem with vandalism/graffiti	£4,072	£2,443	£4,684	£4,906	£3,089	£2,443	£2,443	£6,515	£4,289	£2,443	£5,263	£4,115
ENV1403	» Not worried about crime	£12,274	£17,356	£12,435	£13,978	£19,399	£7,525	£19,399	£16,527	£11,535	£18,813	£11,222	£10,348
ENV1604	» No problem with anti-social behaviour	£6,403	£3,842	£7,199	£6,496	£3,842	£3,842	£5,579	£7,580	£7,057	£4,289	£7,868	£6,310
ENV1605	» Police do good job	£5,340	£6,263	£6,082	£3,204	£8,543	£8,148	£8,543	£3,204	£4,800	£5,959	£5,277	£3,204
ENV1606	» No litter problems	£3,555	£3,173	£3,684	£3,300	£2,133	£2,133	£3,470	£2,133	£3,942	£5,533	£3,813	£3,513
ENV1607	» Able to obtain advice locally	£2,457	£2,507	£1,567	£3,561	£3,919	£3,931	£3,003	£3,931	£1,977	£1,474	£1,474	£3,931
ENV1608	» Good neighbourhood	£1,747	£1,048	£1,048	£2,795	£1,048	£1,048	£1,048	£2,795	£2,054	£1,048	£1,048	£2,795
ENV1609	» Feel belonging to neighbourhood	£3,753	£2,706	£2,252	£6,004	£2,252	£2,252	£2,682	£2,407	£3,919	£3,223	£2,252	£6,004
ENV1410	» Talks to neighbours regularly	£4,511	£3,369	£3,195	£6,820	£3,910	£2,972	£4,007	£6,820	£4,535	£3,837	£3,070	£5,075
HEA1601	⁘ High confidence (adult)	£13,080	£14,224	£13,065	£12,565	£13,188	£15,264	£12,801	£12,817	£13,065	£14,152	£13,096	£12,549
HEA1602	⁘ Relief from depression/anxiety (adult)	£36,766	£33,107	£36,949	£38,800	£35,563	£52,293	£38,053	£28,627	£36,827	£31,914	£36,706	£39,302
HEA1603	⁘ Good overall health	£20,141	£16,921	£21,200	£20,323	£23,338	£25,869	£24,282	£21,715	£19,913	£16,412	£20,922	£20,186
HEA1604	⁘ Relief from drug/alcohol problems	£26,124	£30,633	£30,688	£15,674	£41,798	£41,798	£41,798	£29,324	£24,120	£25,616	£29,540	£15,674
HEA1605	⁘ Smoking cessation	£4,010	£4,571	£4,072	£3,856	£2,410	£6,416	£2,406	£2,406	£4,041	£4,414	£4,196	£3,948
HEA1406	⁘ Feel in control of life	£15,894	£15,637	£17,504	£16,427	£14,894	£18,607	£12,833	£15,311	£15,878	£14,399	£16,474	£15,734
HEA1607	⁘ Can rely on family	£6,784	£10,855	£6,784	£6,742	£7,532	£10,855	£7,876	£4,389	£6,700	£10,855	£6,636	E7,036

Figure 7.12 The Housing Associations' Charitable Trust (HACT) Social Value Bank (2018).

Global Value Exchange (GVE)

The Global Value Exchange is an online portal for social change metrics which includes work resulting from hundreds of organisations and researchers (Bichard, 2015) and includes a wide variety of databases, including the English Indices of Deprivation, the Happy City Index (measures wellbeing valuation), the Progress out of Poverty Index, The Journey to Employment Framework, the Economy for the Common Good Balance Sheet, the New Economy Unit Cost Database, the NICE Indicators, the Global Youth Development Index and the NHS Outcomes Framework. All indicators and proxies can be searched by inputting key words into the search facility on their web site.

The Economics of Ecosystems and Biodiversity (TEEB) Valuation Database

The TEEB database includes a large database on monetary values of ecosystem services which now contains over 1,350 data points from over 300 case studies (GVE, 2017). After the release of the TEEB Valuation Database in 2010, the authors continued development of the database under the name "Ecosystem Services Valuation Database" (ESVD).

New Economy Manchester Unit Cost Database

The New Economy Manchester Unit Cost Database includes indicators and proxies within seven categories, including crime, education and skills, employment and economy, fire, health, housing and social services and provides a very comprehensive set of data for use within the method.

Stage 3 →

at changes)

Indicator	Source	Quantity	Duration	Outcomes start	Financial Proxy	Value in currency	Source
How would you measure it?	Where did you get the information from?	How much change was there?	How long does it last after end of activity? *(Only enter numbers)*	Does it start in period of activity (1) or in period after (2)	What proxy would you use to value the change?	What is the value of the change? *(Only enter numbers)*	Where did you get the information from?

Figure 7.13 Stage three of the SuHousingImpact method.

Government statistics

Government statistics are searchable via the web site of the Office for National Statistics (ONS, 2018), which has access to over 2,000 datasets, including road safety data, crime statistics, databases published by the NHS, information on regeneration areas, housing and social housing (Figure 7.13).

It should be borne in mind that specific stakeholder engagement should be sought, resulting in primary data as far as possible which should be used in conjunction with such information highlighted above. It is not enough to simply use secondary data. Ideally, both primary and secondary data should be used to provide a full stakeholder account of impacts.

Stage four

During stage four of the method, counterfactuals are used before the final impact of a scheme is conveyed. This involves deadweight, referred to as what would have happened anyway – deadweight is always considered to be an estimate due to the fact that a perfect comparison with the same group of people not benefitting from the same intervention is not possible (Nicholls et al., 2012). Next, displacement is an assessment of how much of an outcome

Stage 4 →

Deadweight %	Displacement %	Attribution %	Drop off %	Impact
What would have happened without the activity?	What activity did you displace?	Who else contributed to the change?	Does the outcome drop off in future years?	Quantity times financial proxy, less deadweight,displacement and attribution
0%	0%	0%	0%	0.00
0%	0%	0%	0%	0.00
0%	0%	0%	0%	0.00
0%	0%	0%	0%	0.00
0%	0%	0%	0%	0.00
0%	0%	0%	0%	0.00

Figure 7.14 Stage four of the SuHousingImpact method.

has displaced other outcomes. It does not apply to every analysis but it is however important to be aware of this notion (Nicholls et al., 2012). In addition, its application ensures that any "negative issues that regeneration seeks to eliminate" (Higham et al., 2017) would not simply be transferred to another community, thereby giving a false reading of impact. Attribution is the assessment of how much of an outcome was caused by other organisations or people and at its most basic level involves making sure that all relevant stakeholders are included in the process (Nicholls et al., 2012). The authors state that drop off is "usually calculated by deducting a fixed percentage from the remaining level of outcome at the end of each year" (Ibid, 2012). The counterfactuals adhere to another of the aforementioned SROI principles in not overclaiming (Figure 7.14).

Stage five

Stage five involves the application of a discounted rate to the impact value, again, used in order so as not to over-claim. There are a range of different rates that could be applied to totals within stage five of the process. However, for the public sector, the rate recommended in HM Treasury's Green Book is that of 3.5%. Formulas are embedded within the cells in this stage of the impact map to calculate the discounted values (Figure 7.15).

Stage 5					
Calculating Social Return					
Discount rate		3.5%			
Year 0	Year 1	Year 2	Year 3	Year 4	Year 5
0.00	0.00	0.00	0.00	0.00	0.00
0.00	0.00	0.00	0.00	0.00	0.00
0.00	0.00	0.00	0.00	0.00	0.00
0.00	0.00	0.00	0.00	0.00	0.00
0.00	0.00	0.00	0.00	0.00	0.00
0.00	0.00	0.00	0.00	0.00	0.00
0.00	0.00	0.00	0.00	0.00	0.00
0.00	0.00	0.00	0.00	0.00	0.00
0.00	0.00	0.00	0.00	0.00	0.00

Figure 7.15 Stage five of the SuHousingImpact method.

Stage six

Stage six involves the novel contribution of this tool and is the stage that has refined Bichard's SuROI methodology into the 'SuHousingImpact' tool. This stage splits numerical and quantifiable values within the impact map to show how potential stakeholders can either be winners or losers from a particular housing-led urban regeneration scheme, and by how much, in terms of socio-environmental impact. This stage is particularly useful in terms of being able to evidence how much or how little a stakeholder stands to gain from a particular scheme, which is useful in terms of the aforementioned offloading of costs to various involved stakeholders to ensure continued economic sustainability. In addition, this stage uses these same impact values to create payback periods for each stakeholder and for the scheme as a whole. This is very useful as a strategic decision making, management or governance tool. These calculations, which have never before featured as part of a SuROI process, can be used to prioritise schemes based on how quickly the payback periods come along or alternatively it would be able to potentially cancel plans for any schemes which take too long to break even and it would additionally be able to be used to target certain groups of stakeholders that housing associations or housing-led urban regeneration facilitators wish to target as part of a scheme, either such stakeholders who previously have not benefitted from any such schemes or stakeholders who are currently benefitting from schemes at a cost to the housing association and who could potentially supplement such costs, thus ensuring continued economic sustainability (Figure 7.16).

The workings and function of the payback period table and graph analyses can be seen within the simple examples below. If a stakeholder (such as Housing Association X's customers) is made aware of the evidence-based socio-environmental benefits coming its way through a housing-led urban regeneration scheme, and that stakeholder agrees to pay a capital investment amount towards the financial sustainability of the scheme going forward, to the sum of, for argument's sake, £100, then the payback period can be calculated to be within year 1, and seen within the table and corresponding graph in Figure 7.17.

Similarly, the same can be carried out, with a hypothetical sum of £200, where the payback period would then be within year two, as per Figure 7.18.

The same can be carried out for the capital investment sum of £300, where the payback period is then within year 3, see Figure 7.19.

The same calculation can be made with a capital investment sum of £400, where the payback period would then be within year four, see Figure 7.20.

Finally, the same calculation can be made with a capital investment sum of £500, where the payback period would then be within year five, see Figure 7.21.

Stakeholder A	Cash inflow	Cash outflow	Cumulative cash flow
Y0	0	−100000	−100000
Y1	£ 912,320.24	0	£ 812,320.24
Y2	£ 880,389.03	0	£ 1,692,709.27
Y3	£ 849,575.42	0	£ 2,542,284.69
Y4	£ 819,840.28	0	£ 3,362,124.96
Y5	£ 791,145.87	0	£ 4,153,270.83

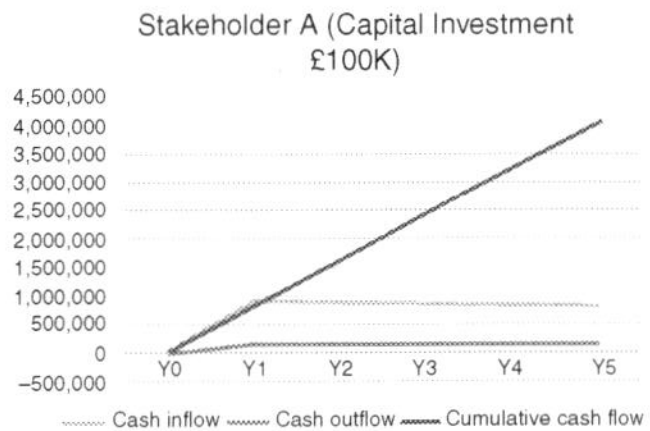

Stakeholder B	Cash inflow	Cash outflow	Cumulative cash flow
Y0	0	−50000	−50000
Y1	£ 312,000.00	0	£ 262,000.00
Y2	£ 301,080.00	0	£ 563,080.00
Y3	£ 290,542.20	0	£ 853,622.20
Y4	£ 280,373.22	0	£ 1,133,995.42
Y5	£ 270,560.16	0	£ 1,404,555.58

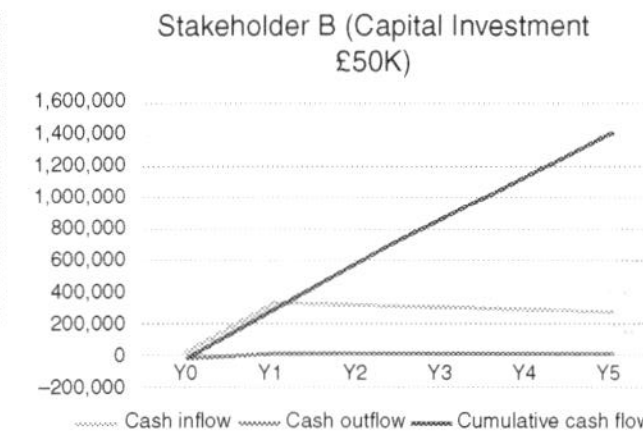

Stakeholder A	Cash inflow	Cash outflow	Cumulative cash flow
Y0	0	−500000	−500000
Y1	£ 912,320.24	0	£ 412,320.24
Y2	£ 880,389.03	0	£ 1,292,709.27
Y3	£ 849,575.42	0	£ 2,142,284.69
Y4	£ 819,840.28	0	£ 2,962,124.96
Y5	£ 791,145.87	0	£ 3,753,270.83

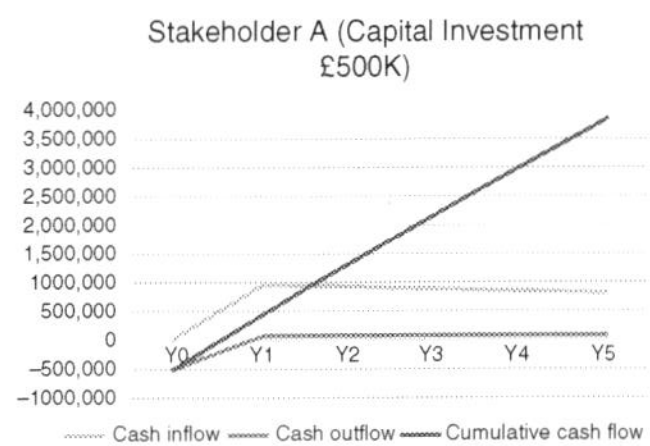

Stakeholder B	Cash inflow	Cash outflow	Cumulative cash flow
Y0	0	−500000	−500000
Y1	£ 312,000.00	0	−£ 188,000.00
Y2	£ 301,080.00	0	£ 113,080.00
Y3	£ 290,542.20	0	£ 403,622.20
Y4	£ 280,373.22	0	£ 683,995.42
Y5	£ 270,560.16	0	£ 954,555.58

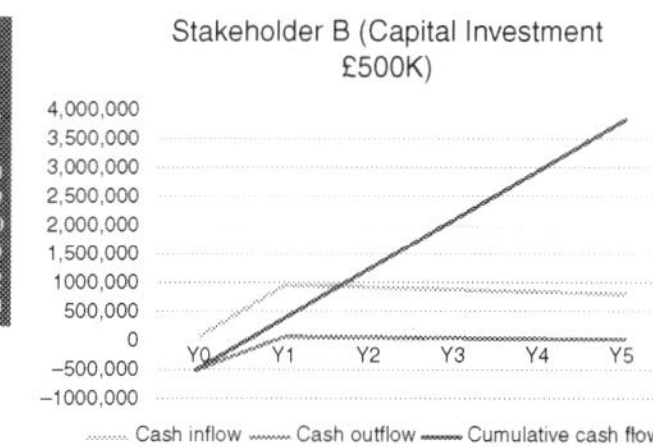

Figure 7.16 Stage six of the SuHousingImpact method.

Stakeholder A	Cash inflow	Cash outflow	Cumulative cash flow
Y0	0	−100	−100.00
Y1	£ 100.00	0	0.00
Y2	£ 100.00	0	£ 100.00
Y3	£ 100.00	0	£ 200.00
Y4	£ 100.00	0	£ 300.00
Y5	£ 100.00	0	£ 400.00

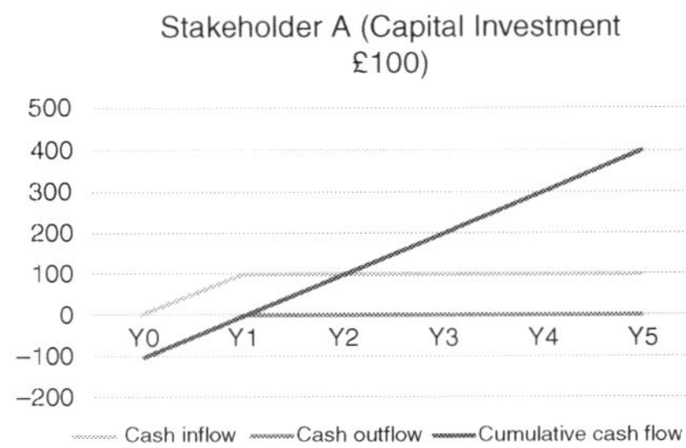

Figure 7.17 Capital investment of £100 payback period hypothetical calculation.

Stakeholder A	Cash inflow	Cash outflow	Cumulative cash flow
Y0	0	−200	−200
Y1	£ 100.00	0	−100.00
Y2	£ 100.00	0	0.00
Y3	£ 100.00	0	£ 100.00
Y4	£ 100.00	0	£ 200.00
Y5	£ 100.00	0	£ 300.00

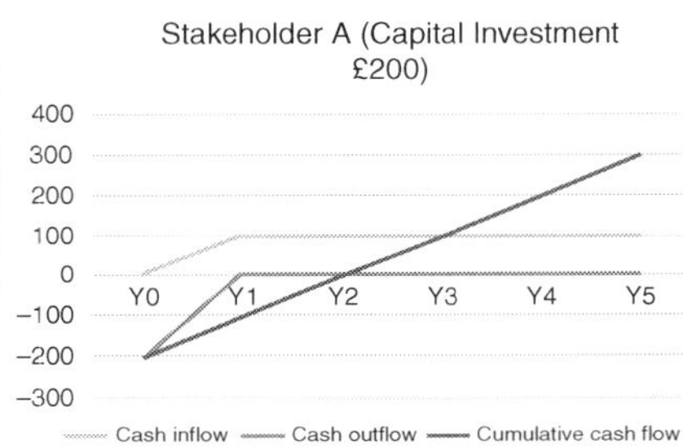

Figure 7.18 Capital investment of £200 payback period hypothetical calculation.

Stakeholder A	Cash inflow	Cash outflow	Cumulative cash flow
Y0	0	−300	−300
Y1	£ 100.00	0	−200.00
Y2	£ 100.00	0	−100.00
Y3	£ 100.00	0	0.00
Y4	£ 100.00	0	£ 100.00
Y5	£ 100.00	0	£ 200.00

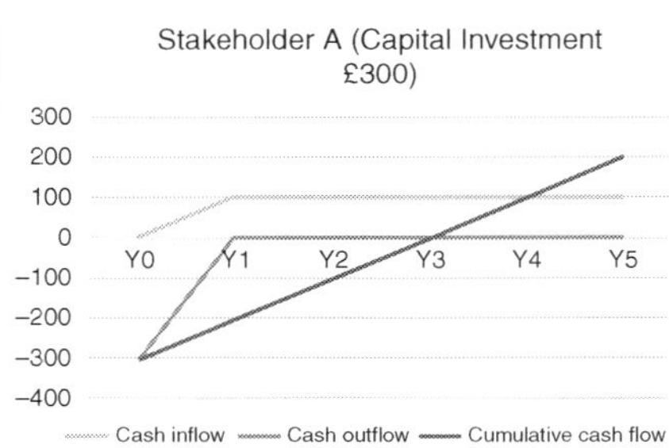

Figure 7.19 Capital investment of £300 payback period hypothetical calculation.

Stakeholder A	Cash inflow	Cash outflow	Cumulative cash flow
Y0	0	−400	−400
Y1	£ 100.00	0	−300.00
Y2	£ 100.00	0	−200.00
Y3	£ 100.00	0	−100.00
Y4	£ 100.00	0	0.00
Y5	£ 100.00	0	£ 100.00

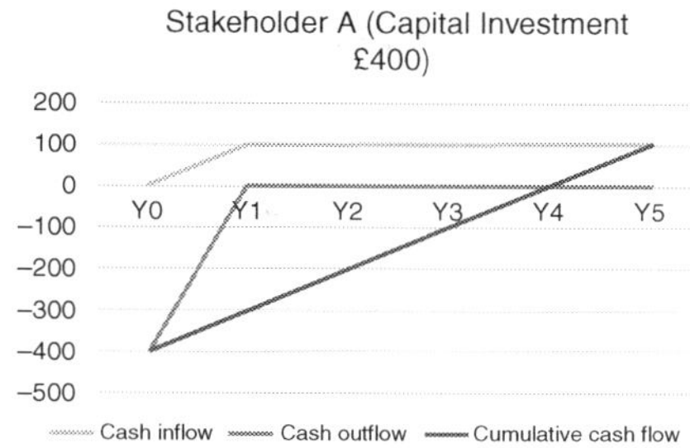

Figure 7.20 Capital investment of £400 payback period hypothetical calculation.

Stakeholder A	Cash inflow	Cash outflow	Cumulative cash flow
Y0	0	−500	−500
Y1	£ 100.00	0	−400.00
Y2	£ 100.00	0	−300.00
Y3	£ 100.00	0	−200.00
Y4	£ 100.00	0	−100.00
Y5	£ 100.00	0	0.00

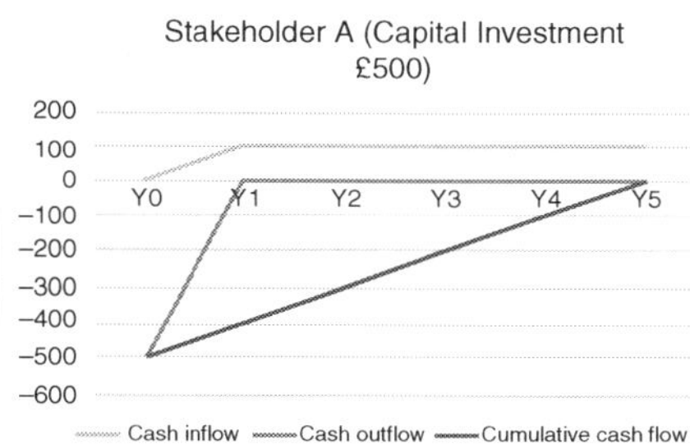

Figure 7.21 Capital investment of £500 payback period hypothetical calculation.

Summary

This chapter explained how the new method (the 'SuHousingImpact' tool), which has been developed, functions. Screenshots of the method together with step-by-step instructions were provided in order that readers of this book, including academics, students, public sector employees, urban regeneration professionals and housing association staff alike, are able to fully understand and subsequently implement the method for their own use. Guidance was also provided on sources of indicators and proxies to feed the tool with all necessary quantifiable information on socio-environmental spillovers from housing-led urban regeneration schemes. The tool itself can be accessed via www.suhousingimpact.com.

8 Case studies

Introduction

This chapter provides two case study exemplars by 'Housing Association X' as a means to demonstrate how the SuHousingImpact tool can be useful for housing-led urban regeneration providers, both in terms of evaluating schemes more sustainably, through a quantifiable methodology, and also, of especial importance in the modern day economic climate faced by such regeneration providers, through its potential application in terms of offsetting economic costs onto potential wider stakeholders and, also, through the additional usage of the SuHousingImpact method as a strategic decision making or management tool through the new found information produced by stage 6 of the tool, which gauges impacts per stakeholder, and which introduces the notion of time, through payback period analyses being able to be carried out both per stakeholder and for a scheme as a whole. The first case study is that of an environmental programme by Housing Association X, and the second is a high-rise scheme – these case studies have purposely been selected to demonstrate the application of SuHousingImpact on different housing project types.

Case study: Housing Association X

This chapter presents the application of the SuHousingImpact method within two case study exemplar projects undertaken by Housing Association X, as a demonstrator as how it can be applied in practice. Housing Association X is a not-for-profit housing association that owns and manages over 14,600 homes in the North West of England. It was established in October 2008, following a stock transfer from the local City Council. Following this, the company has worked in partnership with customers and partners to build sustainable communities where local people are empowered to improve their lives through education, training and enterprise.

Housing Association X is passionate about managing, maintaining and building homes and services that local people need – both now and for the future. Its vision of "improved lives" is underpinned by its Community Impact Strategy.

DOI: 10.1201/9781003043881-8

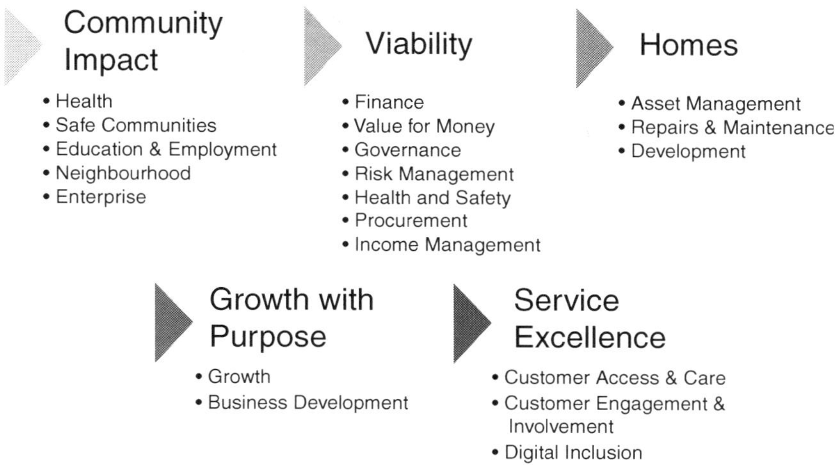

Figure 8.1 Housing Association X's priorities.

This is, about working openly and in partnership so that together it is possible to make a difference to people's lives and futures.

The company has a vision of improved lives which applies right across communities, customers and staff; it values innovation, trust, openness, excellence, respect and passion, and has the following priorities (Figure 8.1):

Almost £250 million has already been invested by Housing Association X in homes and neighbourhoods, and further major projects are underway. However, these projects work against a backdrop of difficult economic times.

The Environmental-Led programme

By utilising the information gained from the first round of interviews carried out with Housing Association X's key employees (the scoping interviews), the second part of each interviewee's transcript contained questions relating to specific information pertaining to it that could be utilised within the prospective completed impact map of the SuHousingImpact method. The thinking behind this was so that both the case study schemes used in this study could be used to carry out an actual evaluation through the usage of the new SuHousingImpact method.

During 2014/15, Housing Association X delivered high-specification environmental improvements to 476 customer homes. These improved the physical appearance of neighbourhoods, enhanced property security and provided off-street parking. Figures 8.2 and 8.3 show photos of before and after the works:

The questions within the interviews were set up to mirror the columns within the impact map of the SuHousingImpact method, with the information provided by the initial scoping interviews to be used to form assessments on the socio-environmental impact of both the environmental-led programme and

Figure 8.2 Housing Association X's environmental-led programme before works.

Figure 8.3 Housing Association X's environmental-led programme after works.

the high-rise scheme. However, in accordance with the new SuHousingImpact method, these monetary values will also be presented within the new 'stage 6' to show the inputs and impacts per stakeholder together with an accompanying line graph per stakeholder showing the payback period. In addition, the same is carried out for the scheme as a whole.

If the completed impact map of the environmental-led scheme is followed through the stages, from left to right, we can see the following:

The Stakeholders

Involved stakeholders listed by interviewees within the said scoping interviews included Housing Association X's customers, private owners on the estate (the estate within which works were carried out), Housing Association X customers' families, friends or visitors to the estates, the community/public in general, City Council highways department, the involved architect with the scheme, utility companies, contractors, Housing Association X and its staff, police teams who deal with low-level issues such as parking issues and Healthcare Provider Y.

Stakeholder details are inputted into stage 1. Also inputted into stage 1 are any intended or unintended changes resulting from the said scheme.

Interviewees came up with different changes that they thought were applicable which were then subsequently entered into the impact map. Such intended or unintended changes involved per stakeholder included:

- Housing Association X Customers:
  Improved security
  Improved parking provision and increased safety
  Aesthetics improved, including better and more uniformed appearance/environment
  Happiness/wellbeing/pride/quality of life/customer satisfaction
  Safety, including traffic safety
  Lower maintenance levels
  Fewer arguments over parking
- Private owners on the estate:
  Improved parking provision
  Improved aesthetics of area/better environment
  Value of property
  Potential improvement of ASB (anti-social behaviour)/crime
  Improved wellbeing/pride/quality of life
  Traffic safety improvement
- Housing Association X customers' families, friends or visitors to the estate
  Improved parking provision and congestion reduction
  Improved aesthetics of area
  Reduction in crime
  Traffic safety improvement
  Regeneration impact to local area

- Community/public in general
  Improved parking provision and congestion reduction
  Improved aesthetics of area
  Reduction in crime
  Traffic safety improvement
  Regeneration impact to local area
- City Council highways department
  Improvements to pavements (dropped kerbs, to enable off-road vehicle parking)
- Architect
  The architect drew up plans which were integral to the scheme
- Utility companies
  Utility information supplied pre scheme
- Contractors
  Contractors' staff working directly on the scheme
- Housing Association X and its staff
  Increased value of stock
  Sustainability and regeneration impact
  Fewer complaints
  Lower maintenance costs
  Reduction in crime and ASB (anti-social behaviour)
  Better void turnover and thereby, rental income
  Customers take more ownership/care more
  Investment into area by Housing Trust through environmental scheme
  Staff involvement and related costs
- Police team dealing with such aspects as parking issues
  Fewer issues to deal with by organisations such as the Police
- Healthcare Provider Y
  Health benefits brought about by the scheme for the local population

Once stage 1 is complete, then the evaluation moves onto stage 2. Stage 2 involves the inputting of information belonging to categories, such as inputs (which can be time or money for example) and also outputs and outcomes.

For the environmental-led scheme, inputs included the price of the scheme itself by way of cost value (£3,200,000), followed by £50,000 worth of Housing Association X officers' staff involvement costs.

The input costs in total are calculated to be £3,250,000 in this case.

Summaries of activities carried out in numbers, also inputted within stage 2 of the impact map, include the 467 properties subject to works being carried out, and the 300 driveways that were installed.

Stage 3 involves the inputting of numbers to mark the quantified change and in addition to this, appropriate indicator or proxy values. Such examples of the calculations within this stage include:

- The extrapolated total of 111 people (from raw data of 26 people out of 111, extrapolated up to a total of 496, used as a conservative total of one person affected per property) who described positive security change in customer interviews carried out by Housing Association X multiplied by the proxy entitled "savings from reduced burglaries per incident" (GVE, 2017) which results in a total of £1,361 multiplied by 111 persons, which equals a total impact of £151,071.
- The 300 cars taken off the road and thereby improving the safety aspect of the neighbourhood multiplied by the proxy "living in a safe area" (£650 per person per year; HACT, 2018).
- A conservative amount of 10% of persons finding that the environmental scheme made them happier than previously about their surrounding environment (48 multiplied by the proxy value of "good neighbourhood" – £499.38 per person per year (HACT, 2018)).

The impact map then calculates the total impacts in monetary form, as has previously been carried out by SROI and SuROI. This then created a total impact value (taking into account depreciative values) of £10,991,052.30 which produced a net present value figure, after input costs are subtracted, of £7,741,052.29.

The total impact ratio for the scheme was then calculated as: £1: £3.38.

This translates to there being £3.38 worth of socio-environmental impact for every £1 spent on the environmental-led scheme at Housing Association X.

The figures are then manipulated as part of the new 'stage 6' via formulas to provide input and impact values per individual stakeholder. A drop off of 3.5% is calculated as part of the formulas inputted to be taken off the impact total per stakeholder for each year. The corresponding tables with their accompanying line graphs showing the payback period relating to different amounts of capital investment for each separate stakeholder (and the scheme as a whole) are additionally shown in Figures 8.4–8.9.

The tables and line graphs in Figure 8.10 are presented within the actual live SuHousingImpact method.

The final tab within the spreadsheet, entitled 'scheme summary', presents a mini table and line graph for the scheme as a whole, see Figure 8.11.

As can be seen from the above impact calculations, the environmental-led scheme has been calculated at offering a total of £5,332,685.29 worth of social and environmental value, which then drops off by 3.5% every year. Different amounts of capital investment can be added and tested out to gauge the respective payback periods.

Stakeholder A	Cash inflow	Cash outflow	Cumulative cash flow
Y0	0	−100000	−100000
Y1	£ 912,320.24	0	£ 812,320.24
Y2	£ 880,389.03	0	£ 1,692,709.27
Y3	£ 849,575.42	0	£ 2,542,284.69
Y4	£ 819,840.28	0	£ 3,362,124.96
Y5	£ 791,145.87	0	£ 4,153,270.83

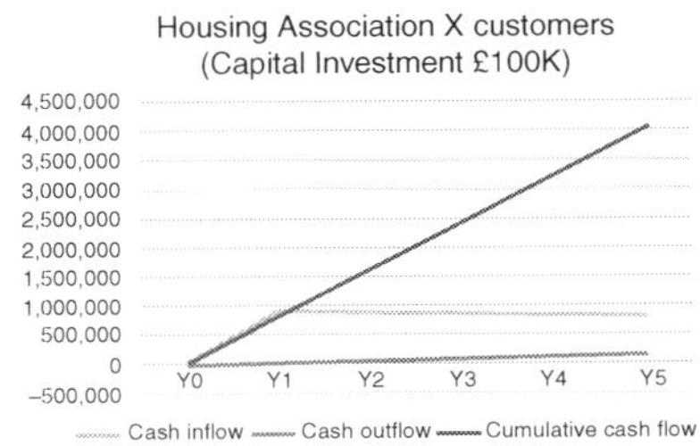

Stakeholder A	Cash inflow	Cash outflow	Cumulative cash flow
Y0	0	−500000	−500000
Y1	£ 912,320.24	0	£ 412,320.24
Y2	£ 880,389.03	0	£ 1,292,709.27
Y3	£ 849,575.42	0	£ 2,142,284.69
Y4	£ 819,840.28	0	£ 2,962,124.96
Y5	£ 791,145.87	0	£ 3,753,270.83

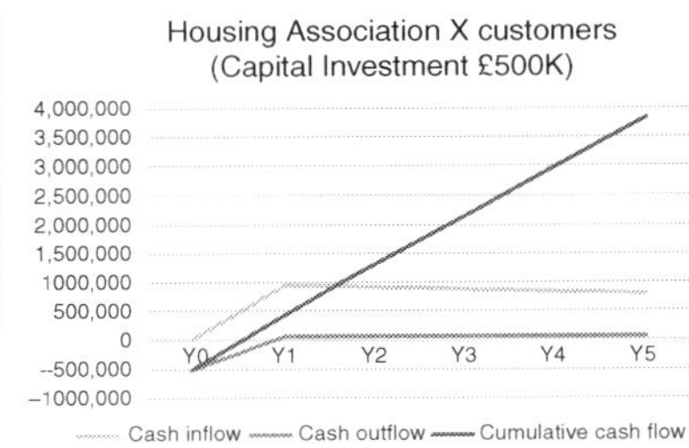

Figure 8.4 Tables and accompanying payback period line graphs for each individually involved stakeholder with different capital investment amounts – 'Housing Association X customers'.

Community/ Public in general	Cash inflow	Cash outflow	Cumulative cash flow
Y0	0	−50000	−50000
Y1	£ 312,000.00	0	£ 262,000.00
Y2	£ 301,080.00	0	£ 563,080.00
Y3	£ 290,542.20	0	£ 853,622.20
Y4	£ 280,373.22	0	£ 1,133,995.42
Y5	£ 270,560.16	0	£ 1,404,555.58

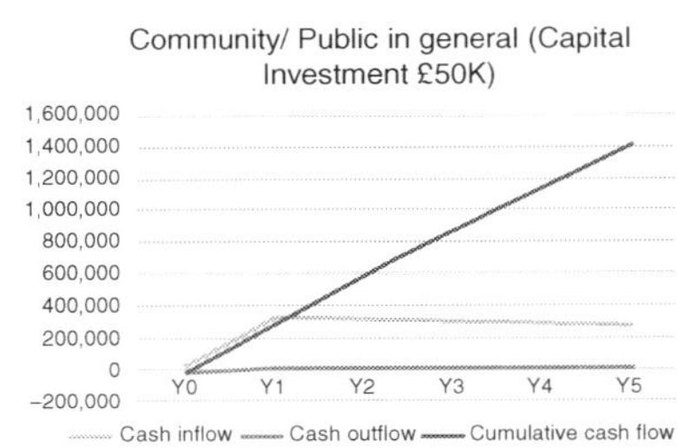

Community/ Public in general	Cash inflow	Cash outflow	Cumulative cash flow
Y0	0	−500000	−500000
Y1	£ 312,000.00	0	£ 188,000.00
Y2	£ 301,080.00	0	£ 113,080.00
Y3	£ 290,542.20	0	£ 403,622.20
Y4	£ 280,373.22	0	£ 683,995.42
Y5	£ 270,560.16	0	£ 954,555.58

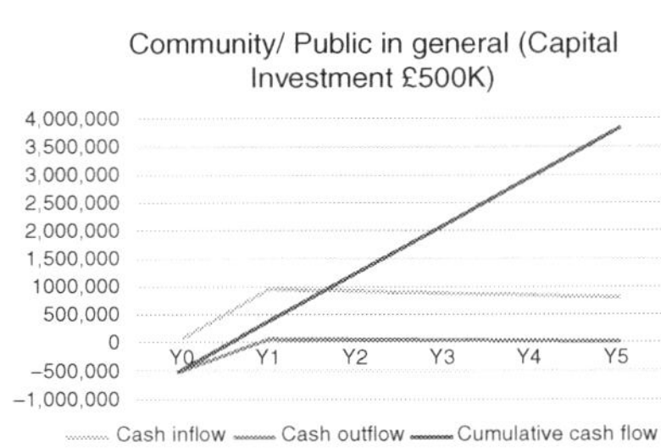

Figure 8.5 Tables and accompanying payback period line graphs for each individually involved stakeholder with different capital investment amounts – 'Community/Public in general'.

City Council highways department	Cash inflow	Cash outflow	Cumulative cash flow
Y0	0	−500	−500
Y1	£ 1,650.00	0	£ 1,150.00
Y2	£ 1,592.25	0	£ 2,742.25
Y3	£ 1,536.52	0	£ 4,278.77
Y4	£ 1,482.74	0	£ 5,761.51
Y5	£ 1,430.85	0	£ 7,192.36

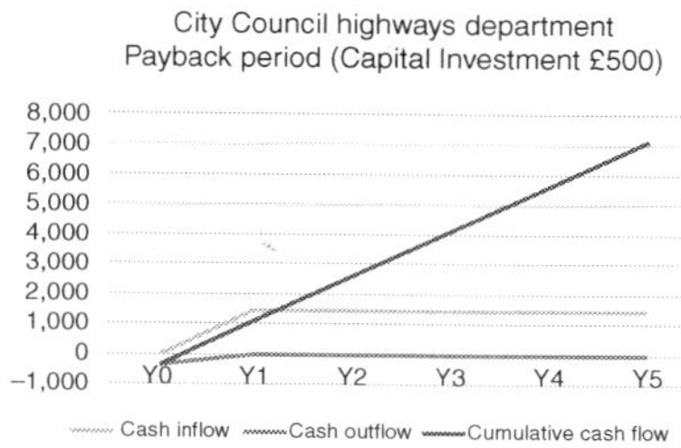

City Council highways department	Cash inflow	Cash outflow	Cumulative cash flow
Y0	0	−1000	−1000
Y1	£ 1,650.00	0	£ 650.00
Y2	£ 1,592.25	0	£ 2,242.25
Y3	£ 1,536.52	0	£ 3,778.77
Y4	£ 1,482.74	0	£ 5,261.51
Y5	£ 1,430.85	0	£ 6,692.36

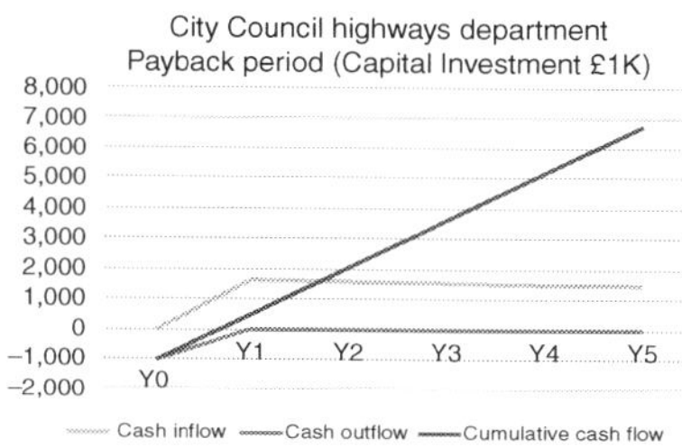

Figure 8.6 Tables and accompanying payback period line graphs for each individually involved stakeholder with different capital investment amounts – 'City Council highways department'.

Housing Association X and its staff	Cash inflow	Cash outflow	Cumulative cash flow
Y0	0	−250000	−250000
Y1	£ 3,139,795.00	0	£ 2,889,795.00
Y2	£ 3,029,902.18	0	£ 5,919,697.18
Y3	£ 2,923,855.60	0	£ 8,843,552.77
Y4	£ 2,821,520.65	0	£ 11,665,073.43
Y5	£ 2,722,767.43	0	£ 14,387,840.86

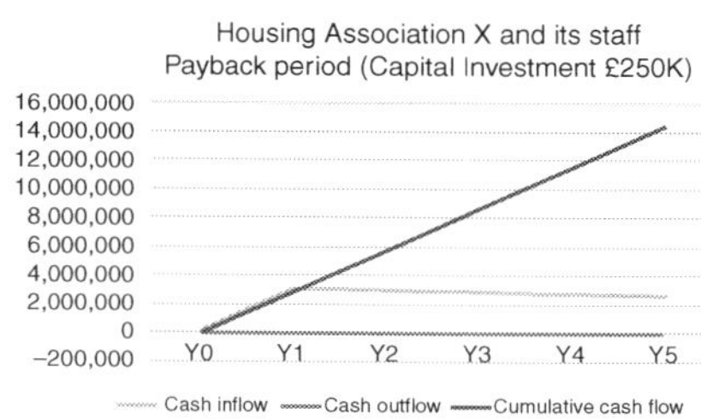

Housing Association X and its staff	Cash inflow	Cash outflow	Cumulative cash flow
Y0	0	−1000	−1000
Y1	£ 3,139,795.00	0	£ 3,138,795.00
Y2	£ 3,029,902.18	0	£ 6,168,697.18
Y3	£ 2,923,855.60	0	£ 9,092,552.77
Y4	£ 2,821,520.65	0	£ 11,914,073.43
Y5	£ 2,722,767.43	0	£ 14,636,840.86

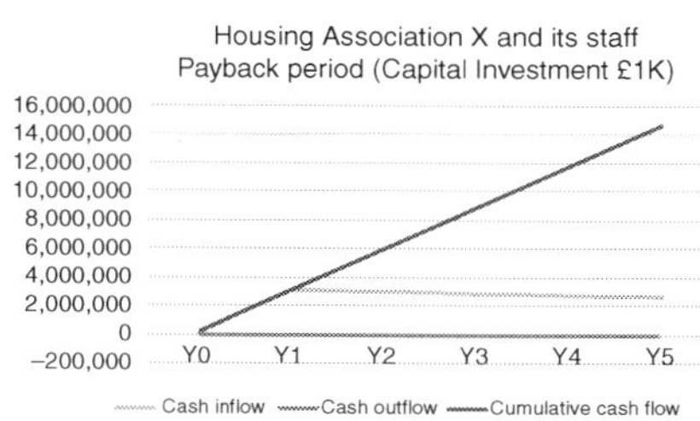

Figure 8.7 Tables and accompanying payback period line graphs for each individually involved stakeholder with different capital investment amounts – 'Housing Association X and its staff'.

Police team dealing with such aspects as parking issues	Cash inflow	Cash outflow	Cumulative cash flow
Y0	0	−100	−100
Y1	£ 152.05	0	£ 52.05
Y2	£ 146.73	0	£ 198.78
Y3	£ 141.59	0	£ 340.37
Y4	£ 136.64	0	£ 477.01
Y5	£ 131.85	0	£ 608.86

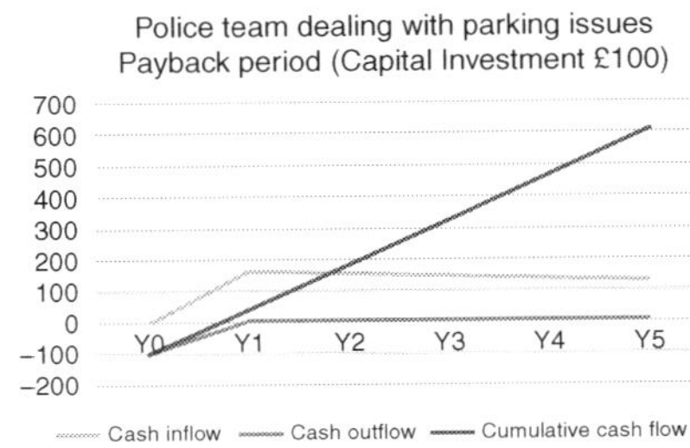

Police team dealing with such aspects as parking issues	Cash inflow	Cash outflow	Cumulative cash flow
Y0	0	−500	−500
Y1	£ 152.05	0	−£ 347.95
Y2	£ 146.73	0	−£ 201.22
Y3	£ 141.59	0	−£ 59.63
Y4	£ 136.64	0	£ 77.01
Y5	£ 131.85	0	£ 208.86

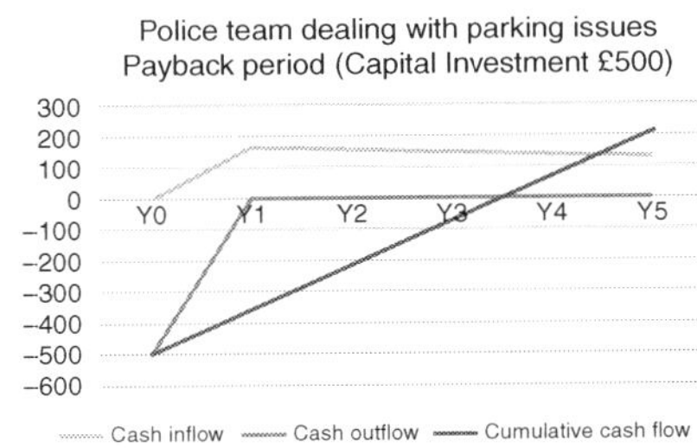

Figure 8.8 Tables and accompanying payback period line graphs for each individually involved stakeholder with different capital investment amounts – 'Police team dealing with parking issues'.

Healthcare Provider Y	Cash inflow	Cash outflow	Cumulative cash flow
Y0	0	−25000	−25000
Y1	£ 966,768.00	0	£ 941,768.00
Y2	£ 932,931.12	0	£ 1,874,699.12
Y3	£ 900,278.53	0	£ 2,774,977.65
Y4	£ 868,768.78	0	£ 3,643,746.43
Y5	£ 838,361.87	0	£ 4,482,108.31

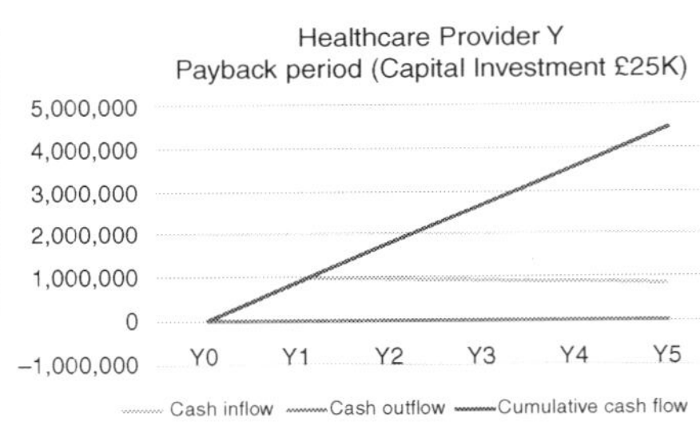

Healthcare Provider Y	Cash inflow	Cash outflow	Cumulative cash flow
Y0	0	−250000	−250000
Y1	£ 966,768.00	0	£ 716,768.00
Y2	£ 932,931.12	0	£ 1,649,699.12
Y3	£ 900,278.53	0	£ 2,549,977.65
Y4	£ 868,768.78	0	£ 3,418,746.43
Y5	£ 838,361.87	0	£ 4,257,108.31

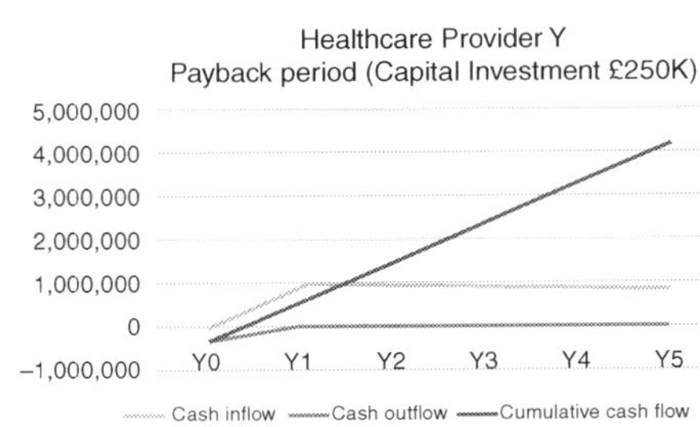

Figure 8.9 Tables and accompanying payback period line graphs for each individually involved stakeholder with different capital investment amounts – 'Healthcare Provider Y'.

Stakeholder A	Cash inflow	Cash outflow	Cumulative cash flow
Y0	0	−100000	−100000
Y1	£ 912,320.24	0	£ 812,320.24
Y2	£ 880,389.03	0	£ 1,692,709.27
Y3	£ 849,575.42	0	£ 2,542,284.69
Y4	£ 819,840.28	0	£ 3,362,124.96
Y5	£ 791,145.87	0	£ 4,153,270.83

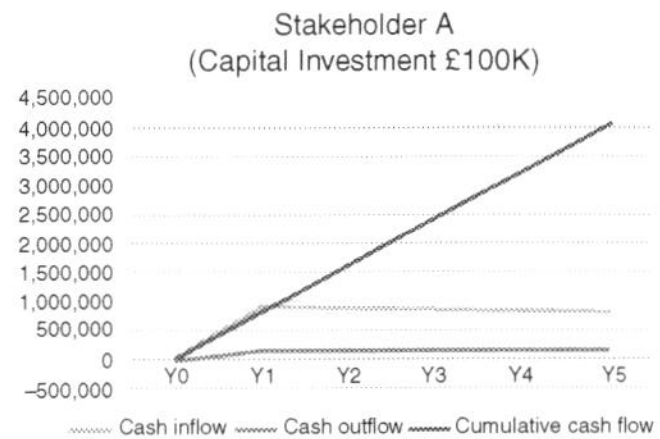

Stakeholder B	Cash inflow	Cash outflow	Cumulative cash flow
Y0	0	−50000	−50000
Y1	£ 312,000.00	0	£ 262,000.00
Y2	£ 301,080.00	0	£ 563,080.00
Y3	£ 290,542.20	0	£ 853,622.20
Y4	£ 280,373.22	0	£ 1,133,995.42
Y5	£ 270,560.16	0	£ 1,404,555.58

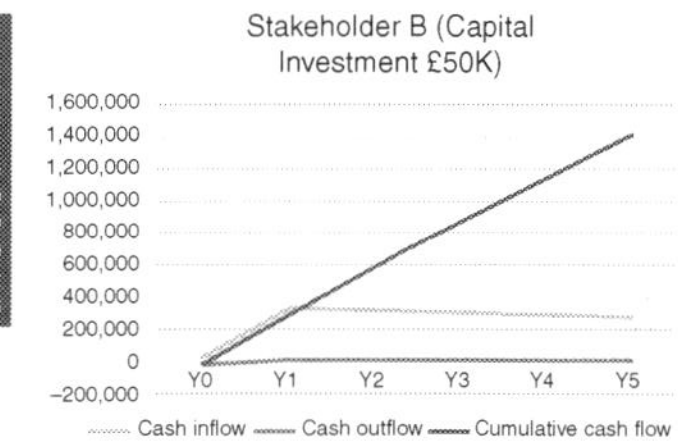

Stakeholder A	Cash inflow	Cash outflow	Cumulative cash flow
Y0	0	−500000	−500000
Y1	£ 912,320.24	0	£ 412,320.24
Y2	£ 880,389.03	0	£ 1,292,709.27
Y3	£ 849,575.42	0	£ 2,142,284.69
Y4	£ 819,840.28	0	£ 2,962,124.96
Y5	£ 791,145.87	0	£ 3,753,270.83

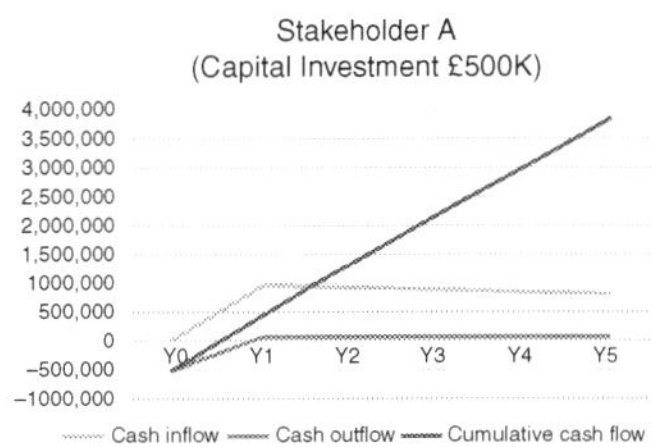

Stakeholder B	Cash inflow	Cash outflow	Cumulative cash flow
Y0	0	−500000	−500000
Y1	£ 312,000.00	0	£ 188,000.00
Y2	£ 301,080.00	0	£ 113,080.00
Y3	£ 290,542.20	0	£ 403,622.20
Y4	£ 280,373.22	0	£ 683,995.42
Y5	£ 270,560.16	0	£ 954,555.58

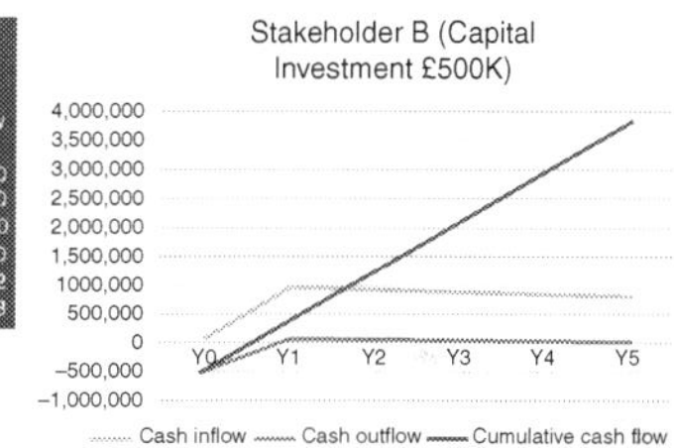

Figure 8.10 Tables and accompanying payback period line graphs for each individually involved stakeholder with different capital investment amounts – live SuHousingImpact method screenshot.

Whole scheme	Cash inflow	Cash outflow	Cumulative cash flow
Y0	0	−5000000	−5000000
Y1	£ 5,332,685.29	0	£ 332,685.29
Y2	£ 5,146,041.30	0	£ 5,478,726.59
Y3	£ 4,965,929.86	0	£ 10,444,656.45
Y4	£ 4,792,122.31	0	£ 15,236,778.77
Y5	£ 4,624,398.03	0	£ 19,861,176.80

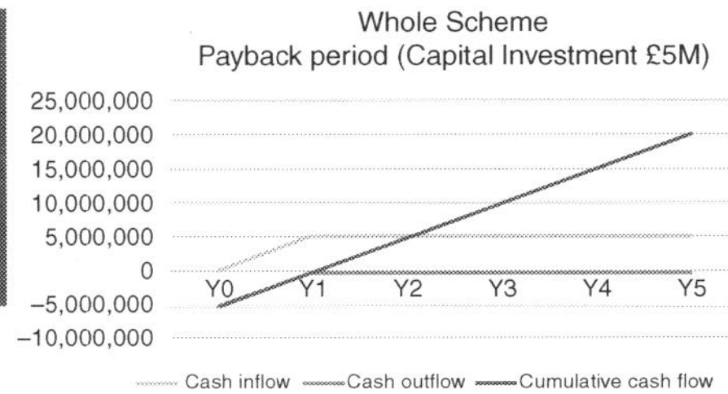

Whole scheme	Cash inflow	Cash outflow	Cumulative cash flow
Y0	0	−10000000	−10000000
Y1	£ 5,332,685.29	0	£ 4,667,314.71
Y2	£ 5,146,041.30	0	£ 478,726.59
Y3	£ 4,965,929.86	0	£ 5,444,656.45
Y4	£ 4,792,122.31	0	£ 10,236,778.77
Y5	£ 4,624,398.03	0	£ 14,861,176.80

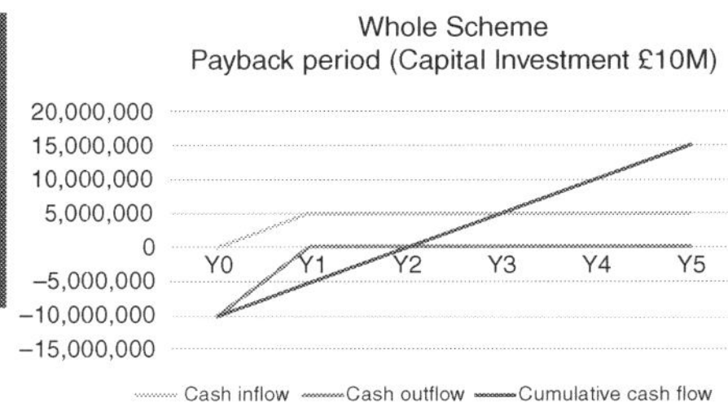

Figure 8.11 Tables and accompanying payback period line graphs for the scheme as a whole.

The High Rise Scheme

Case study 2 involves a high-rise scheme. Housing Association X has invested £43.2m to improve 666 high-rise flats across nine blocks. Improvements include thermal cladding, enclosed conservatory balconies, self-cleaning windows and new lifts, whilst internal improvements include re-modelling to provide open plan living spaces as well as new kitchens, bathrooms, security doors and heating and ventilation systems. Figures 8.12 and 8.13 show before and after photos of the high-rise blocks:

The Housing Association X's high-rise scheme was described as having involved eight stakeholders according to responses within the initial scoping interviews.

The stakeholders listed were as follows:

Housing Association X customers, customers' families, local residents/community/general public, the local authority, including planning, building control and area regeneration, Housing Association X, local businesses, leaseholders and Healthcare Provider Y.

Intended and unintended changes for each individual stakeholder were recounted as being the following:

- Housing Association X customers
  No repairs needed (including bin chute costs)
  Better security and safety

Figure 8.12 Housing Association X high-rise blocks before works.

Figure 8.13 Housing Association X high-rise blocks after works.

Lower ASB (anti-social behaviour)/crime
Blocks attract better people
Improvement aesthetically
Improved pride/feel good factor

Community/social life improved
More desirable place to live
Affordability improved
Quality of life improved
Improved health
Relief from property-related anxiety and embarrassment

- Customers' families
  Less crime/ASB (anti-social behaviour)/better security
  Aesthetic improvement
  Better living standards
  Cost savings for customers impacts potentially on their families also

- Local residents/ community/general public
  Aesthetically improved
  Less ASB (anti-social behaviour)/crime
  More pride
  Better community spirit
  More desirable place to live
  Historic perception of area changed positively

- Council including planning, building control and area regeneration
  More demand within local community for housing, related goods and services and potentially, council tax and/or infrastructure

- Housing Association X
  Reduction in staff costs regarding crime
  More in-demand properties and therefore rent
  Cyclical and call out maintenance less
  More customer ownership
  Cost savings for customers impacts on ability to pay rent

- Local businesses
  Locals have more disposable income
  Area attracts more people to purchase goods

- Leaseholders
  Changes to Housing Association X customers will apply to leaseholders

- Healthcare Provider Y
  Healthcare Provider Y benefits from investment in local area and local population

Once stage 1 is complete, then the evaluation again moves onto stage 2.

Stage 2 involves the inputting of information belonging to categories such as inputs (which can be time or money for example) and also outputs and outcomes. The cost input into the high-rise scheme is the cost of the scheme and included staff costs (£43,200,000).

Associated and relevant output figures include the quantity of 666 high-rise flats involved in the scheme, whilst the sought after outcomes reflect the intended and unintended changes.

Stage 3 incorporates such calculations as the number of high-rise flats affected (666) by a suitable proxy or indicator, in the same way as was carried out for the environmental-led scheme.

For example, regarding the improvement/change of no repairs being needed as a consequence of the improvement works carried out, a calculation of the cost for an average day-to-day repair (£158.56 according to the Housing Association X's Performance Department) was multiplied by the 666 flats to give the impact value of £105,600.96.

Similarly, the impact calculation involving the perception that the neighbourhood had improved, included an extrapolated total of 404 persons (figures extrapolated from interviews from the customer base) multiplied by a relevant proxy value of £1,747 per person for "good neighbourhood", which originates from the HACT (Housing Associations' Charitable Trust) database.

The impact map then calculates the total impacts in monetary form, as has previously been carried out by SROI and SuROI. This then created a total impact value (taking into account depreciative values and minus the input costs) of £38,270,658.85.

The total impact ratio for the scheme was then calculated as: £1: £1.89.

This translates to there being £1.89 worth of socio-environmental impact for every £1 spent on the high-rise scheme at Housing Association X.

The figures are then manipulated as part of the new 'stage 6' via formulas to provide input and impact values per individual stakeholder. A drop off of 3.5% is calculated as part of the formulas inputted to be taken off the impact total per stakeholder for each year. Tables and an accompanying line graph showing the payback period for each separate stakeholder (and the scheme as a whole) are shown in Figures 8.14–8.19.

Again, the live SuHousingImpact method, this time for the high-rise scheme, can be seen within the screenshot, as per Figure 8.20.

Lastly, the mini table and line graph for the high-rise scheme as a whole can be seen in Figure 8.21.

Housing Association X customers	Cash inflow	Cash outflow	Cumulative cash flow
Y0	0	−25000	−25000
Y1	12,612,125.64	0	12,587,125.64
Y2	12,170,701.24	0	24,757,826.88
Y3	11,744,726.70	0	36,502,553.58
Y4	11,333,661.26	0	47,836,214.85
Y5	10,936,983.12	0	58,773,197.97

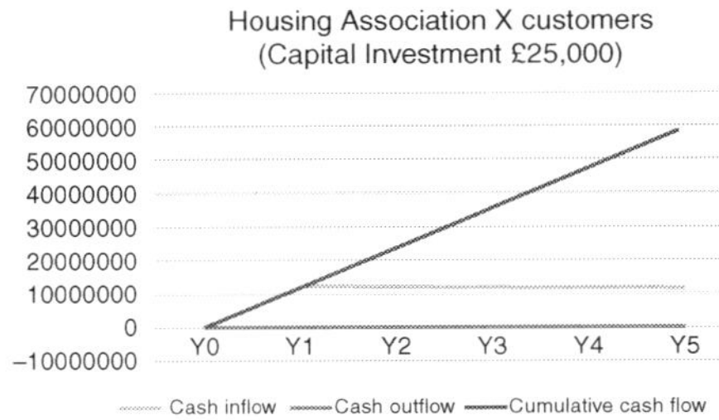

Housing Association X customers	Cash inflow	Cash outflow	Cumulative cash flow
Y0	0	−6000000	−6000000
Y1	12,612,125.64	0	6,612,125.64
Y2	12,170,701.24	0	18,782,826.88
Y3	11,744,726.70	0	30,527,553.58
Y4	11,333,661.26	0	41,861,214.85
Y5	10,936,983.12	0	52,798,197.97

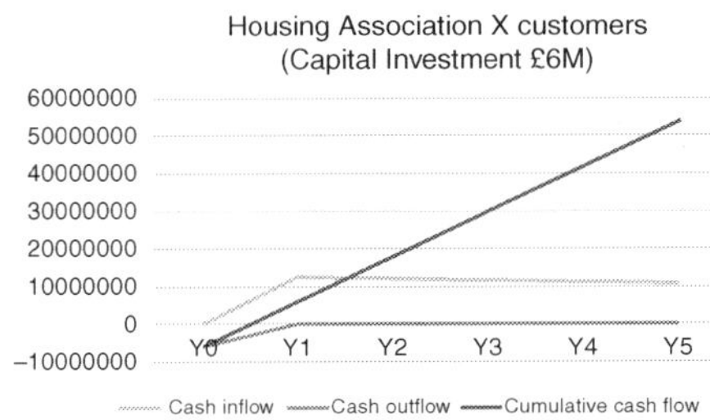

Figure 8.14 Tables and accompanying payback period line graphs for each individually involved stakeholder with different capital investment amounts – 'Housing Association X customers'.

Customers' families	Cash inflow	Cash outflow	Cumulative cash flow
Y0	0	−15,000	−15,000
Y1	33,500.00	0	18,500.00
Y2	32,327.50	0	50,827.50
Y3	31,196.04	0	82,023.54
Y4	30,104.18	0	112,127.71
Y5	29,050.53	0	141,178.24

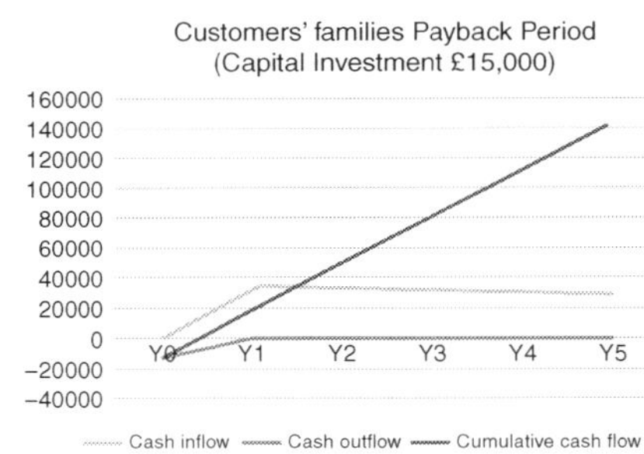

Customers' families	Cash inflow	Cash outflow	Cumulative cash flow
Y0	0	−40,000	−40,000
Y1	33,500.00	0	−6,500.00
Y2	32,327.50	0	25,827.50
Y3	31,196.04	0	57,023.54
Y4	30,104.18	0	87,127.71
Y5	29,050.53	0	116,178.24

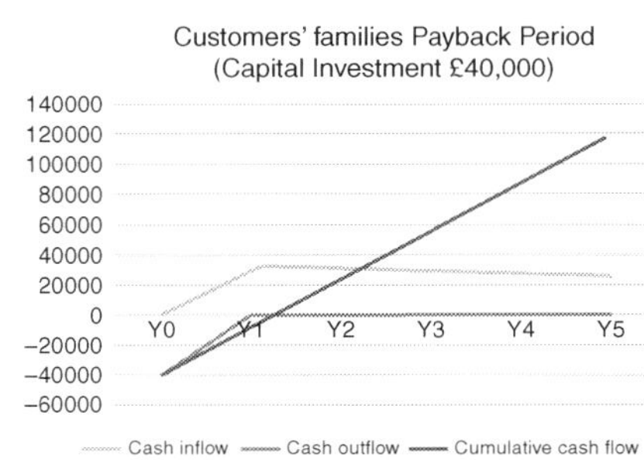

Figure 8.15 Tables and accompanying payback period line graphs for each individually involved stakeholder with different capital investment amounts – 'Customers' families'.

Council	Cash inflow	Cash outflow	Cumulative cash flow
Y0	0	−25,000	−25,000
Y1	4,329,000.00	0	4,304,000.00
Y2	4,177,485.00	0	8,481,485.00
Y3	4,031,273.03	0	12,512,758.03
Y4	3,890,178.47	0	16,402,936.49
Y5	3,754,022.22	0	20,156,958.72

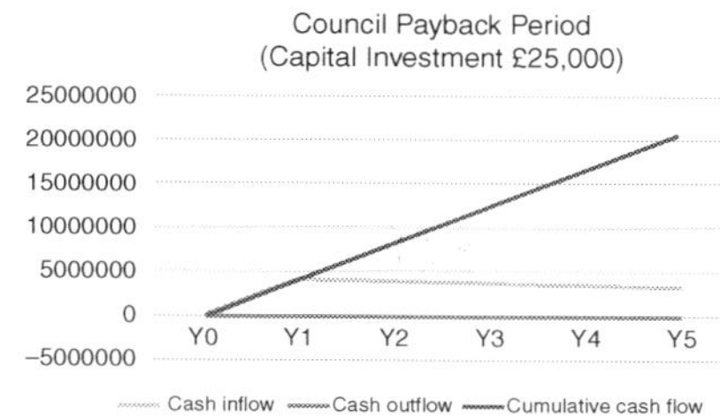

Council	Cash inflow	Cash outflow	Cumulative cash flow
Y0	0	−1,000,000	−1,000,000
Y1	4,329,000.00	0	3,329,000.00
Y2	4,177,485.00	0	7,506,485.00
Y3	4,031,273.03	0	11,537,758.03
Y4	3,890,178.47	0	15,427,936.49
Y5	3,754,022.22	0	19,181,958.72

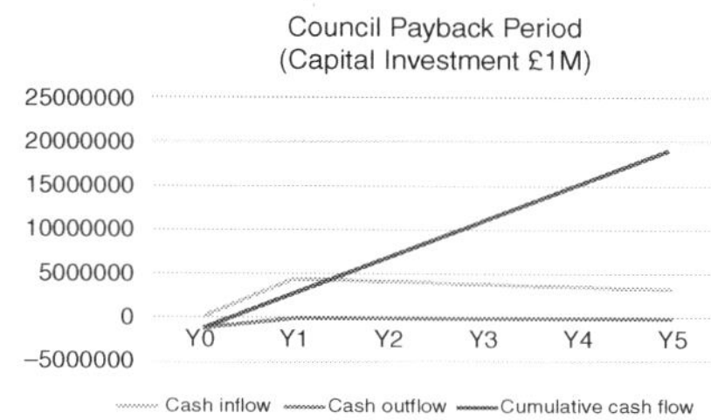

Figure 8.16 Tables and accompanying payback period line graphs for each individually involved stakeholder with different capital investment amounts – 'Council'.

Housing Association X	Cash inflow	Cash outflow	Cumulative cash flow
Y0	0	−25,000	−25,000
Y1	7,937,567.82	0	7,912,567.82
Y2	7,659,752.95	0	15,572,320.77
Y3	7,391,661.59	0	22,963,982.36
Y4	7,132,953.44	0	30,096,935.80
Y5	6,883,300.07	0	36,980,235.72

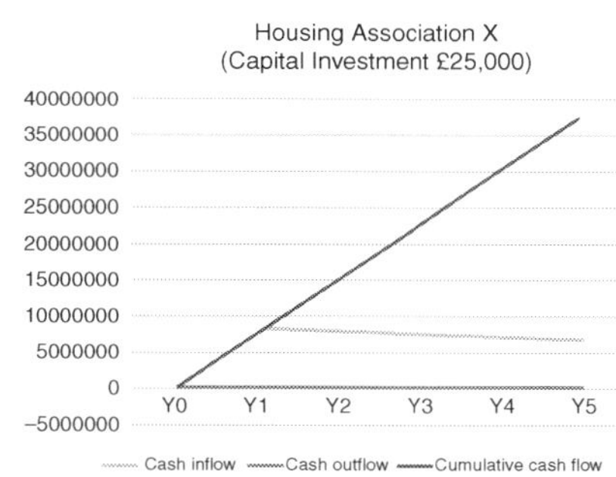

Housing Association X	Cash inflow	Cash outflow	Cumulative cash flow
Y0	0	−5,000,000	−5,000,000
Y1	7,937,567.82	0	2,937,567.82
Y2	7,659,752.95	0	10,597,320.77
Y3	7,391,661.59	0	17,988,982.36
Y4	7,132,953.44	0	25,121,935.80
Y5	6,883,300.07	0	32,005,235.86

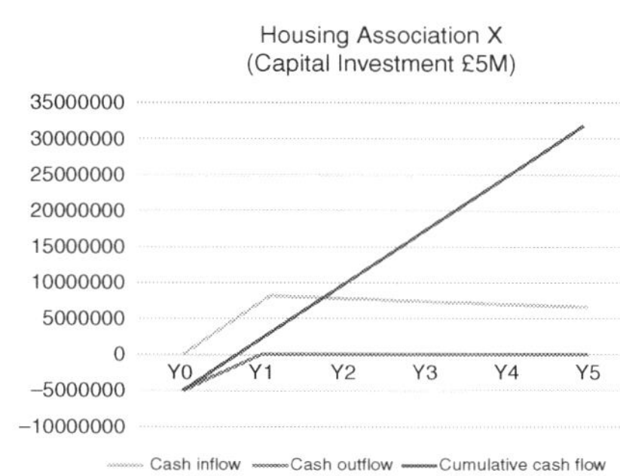

Figure 8.17 Tables and accompanying payback period line graphs for each individually involved stakeholder with different capital investment amounts – 'Housing Association X'.

Leaseholders	Cash inflow	Cash outflow	Cumulative cash flow
Y0	0	−2,000	−2,000
Y1	5,440.00	0	3,440.00
Y2	5,249.60	0	8,689.60
Y3	5,065.86	0	13,755.46
Y4	4,888.56	0	18,644.02
Y5	4,717.46	0	23,361.48

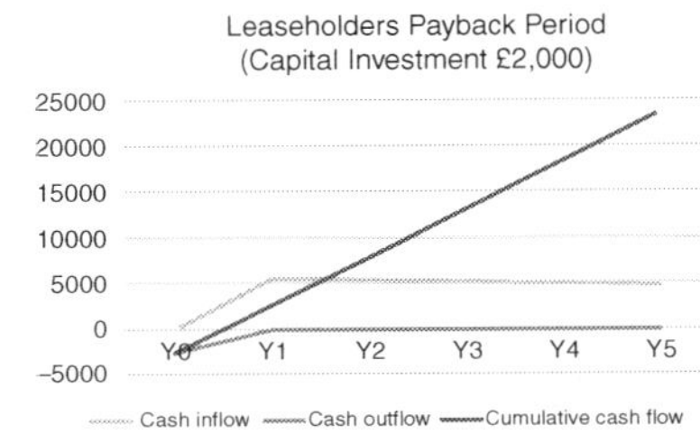

Leaseholders	Cash inflow	Cash outflow	Cumulative cash flow
Y0	0	−10,000	0
Y1	5,440.00	0	5,440.00
Y2	5,249.60	0	10,689.60
Y3	5,065.86	0	15,755.46
Y4	4,888.56	0	20,644.02
Y5	4,717.46	0	25,361.48

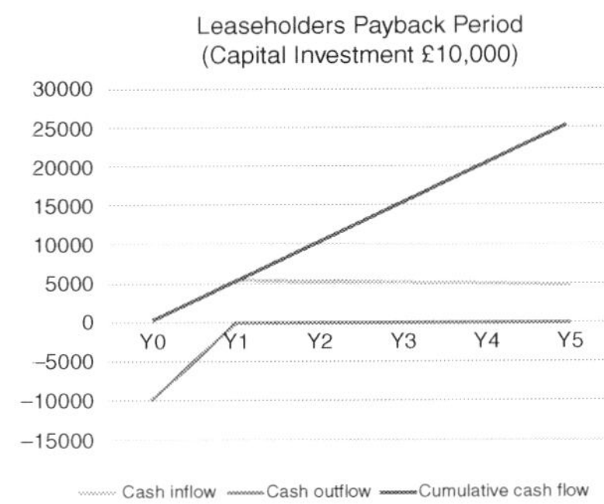

Figure 8.18 Tables and accompanying payback period line graphs for each individually involved stakeholder with different capital investment amounts – 'Leaseholders'.

Healthcare Provider Y	Cash inflow	Cash outflow	Cumulative cash flow
Y0	0	−500	−500
Y1	2,792.00	0	2,292.00
Y2	2,694.28	0	4,986.28
Y3	2,599.98	0	7,586.26
Y4	2,508.98	0	10,095.24
Y5	2,421.17	0	12,516.41

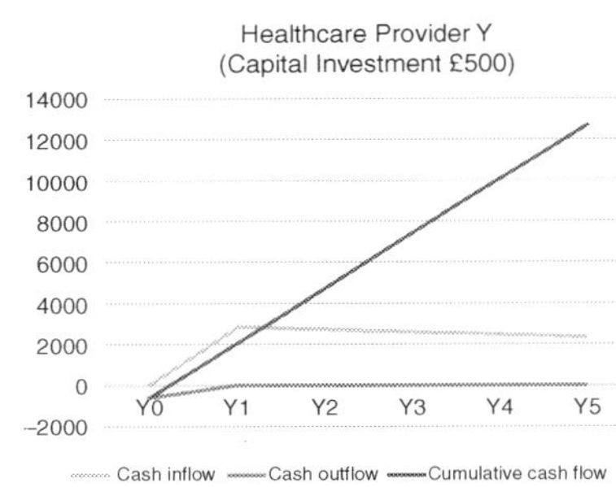

Healthcare Provider Y	Cash inflow	Cash outflow	Cumulative cash flow
Y0	0	−2,000	−2,000
Y1	2,792.00	0	792.00
Y2	2,694.28	0	3,486.28
Y3	2,599.98	0	6,086.26
Y4	2,508.98	0	8,595.24
Y5	2,421.17	0	11,016.41

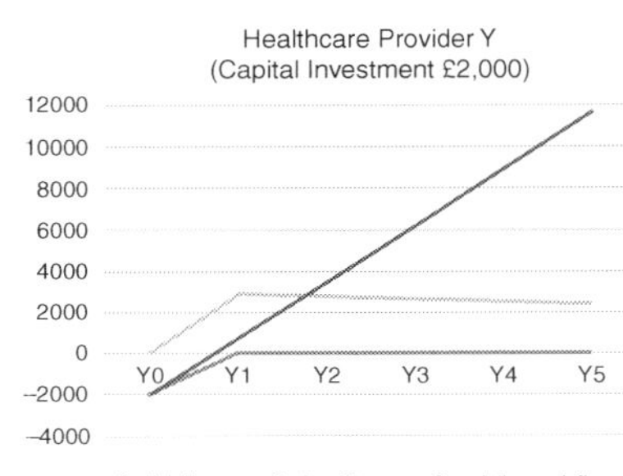

Figure 8.19 Tables and accompanying payback period line graphs for each individually involved stakeholder with different capital investment amounts – 'Healthcare Provider Y'.

Stakeholder A	Cash inflow	Cash outflow	Cumulative cash flow
Y0	0	–25000	–25000
Y1	12,612,125.64	0	12,587,125.64
Y2	12,170,701.24	0	24,757,826.88
Y3	11,744,726.70	0	36,502,553.58
Y4	11,333,661.26	0	47,836,214.85
Y5	10,936,983.12	0	58,773,197.97

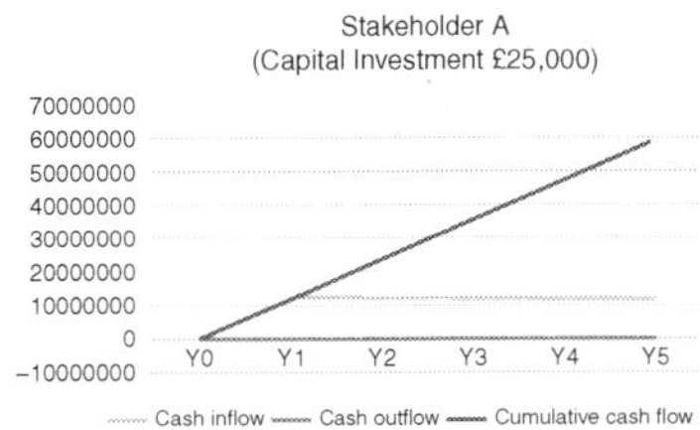

Stakeholder B	Cash inflow	Cash outflow	Cumulative cash flow
Y0	0	–15,000	–15,000
Y1	33,500.00	0	18,500.00
Y2	32,327.50	0	50,827.50
Y3	31,196.04	0	82,023.54
Y4	30,104.18	0	112,127.71
Y5	29,050.53	0	141,178.24

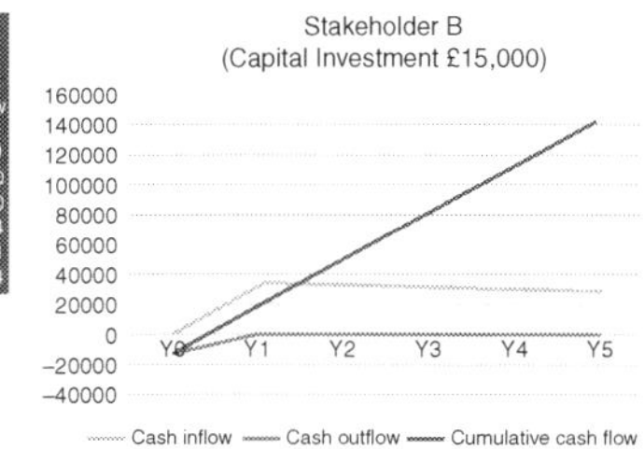

Stakeholder A	Cash inflow	Cash outflow	Cumulative cash flow
Y0	0	–6000000	–6000000
Y1	12,612,125.64	0	6,612,125.64
Y2	12,170,701.24	0	18,782,826.88
Y3	11,744,726.70	0	30,527,553.58
Y4	11,333,661.26	0	41,861,214.85
Y5	10,936,983.12	0	52,798,197.97

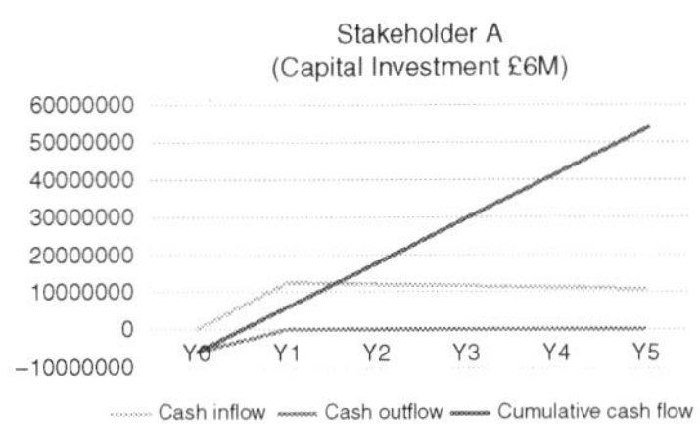

Stakeholder B	Cash inflow	Cash outflow	Cumulative cash flow
Y0	0	–40,000	–40,000
Y1	33,500.00	0	–6,500.00
Y2	32,327.50	0	25,827.50
Y3	31,196.04	0	57,023.54
Y4	30,104.18	0	87,127.71
Y5	29,050.53	0	116,178.24

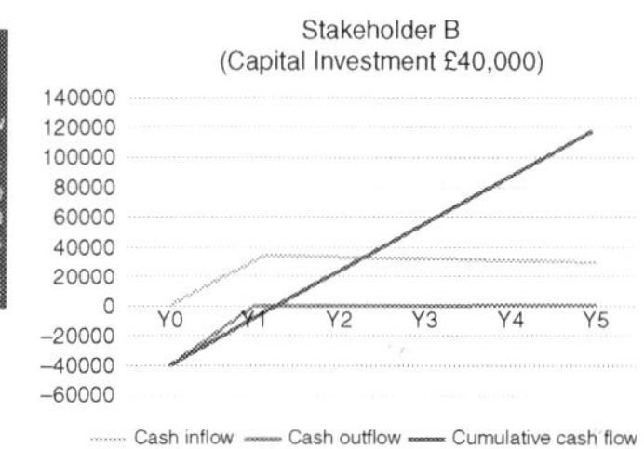

Figure 8.20 Tables and accompanying payback period line graphs for each individually involved stakeholder with different capital investment amounts – live SuHousingImpact method screenshot.

Whole scheme	Cash inflow	Cash outflow	Cumulative cash flow
Y0	0	−5,000,000	−5,000,000
Y1	24,920,425.46	0	19,920,425.46
Y2	24,048,210.57	0	43,968,636.03
Y3	23,206,523.20	0	67,175,159.23
Y4	22,394,294.89	0	89,569,454.11
Y5	21,610,494.57	0	111,179,948.68

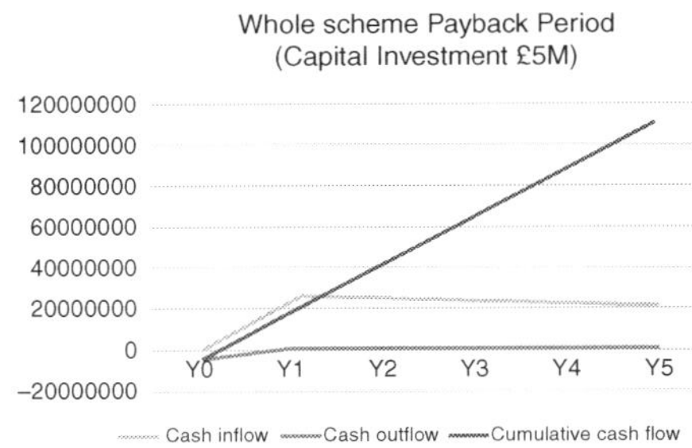

Whole scheme	Cash inflow	Cash outflow	Cumulative cash flow
Y0	0	−10,000,000	−10,000,000
Y1	24,920,425.46	0	14,920,425.46
Y2	24,048,210.57	0	38,968,636.03
Y3	23,206,523.20	0	62,175,159.23
Y4	22,394,294.89	0	84,569,454.11
Y5	21,610,494.57	0	106,179,948.68

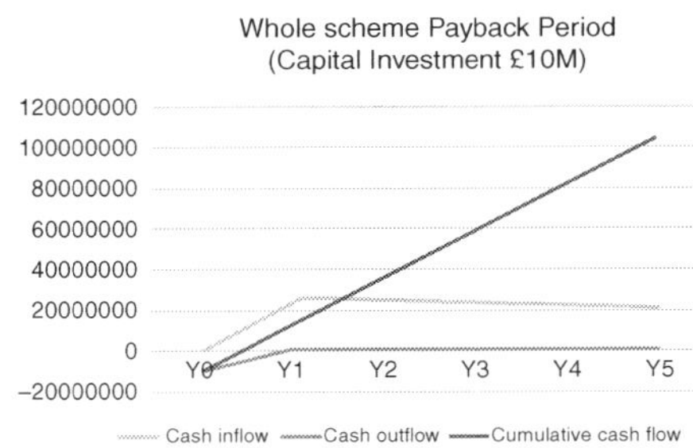

Figure 8.21 Tables and accompanying payback period line graphs for the scheme as a whole.

Summary

This chapter has outlined the SuHousingImpact method, produced as a 'hybrid' or refinement of the SuROI methodology. The tool extends the already existing SuROI methodology by fine tuning the calculations within the impact map spreadsheet to create impact values for each individually involved stakeholder within a housing-led urban regeneration scheme. These values allow users of the tool the ability to view who the winners and losers of a particular scheme are in terms of social and environmental impact. In addition to this, payback period analyses are created as part of the SuHousingImpact method for each individual stakeholder and also for the scheme as a whole. This has created a much needed methodology which not only acts as an evaluating tool, but in addition, one which can be used for strategic decision making, management or governance purposes.

9 Concluding remarks

Conclusion

This book has outlined the method produced as a 'hybrid' or refinement of the SuROI methodology, and has been named the 'SuHousingImpact' tool. The tool extends the already existing SuROI methodology by fine tuning the calculations within the impact map spreadsheet to create impact values for each individually involved stakeholder within a housing-led urban regeneration scheme. These values allow users of the tool the ability to view who the winners and losers of a particular scheme are in terms of social and environmental impact. In addition to this, payback period analyses are created as part of the method for each individual stakeholder and also for the scheme as a whole. This has created a method which not only acts as an evaluation methodology, but in addition, one which can be used for strategic decision making, management or governance purposes.

This book has covered the domains of urban regeneration, sustainable development, evaluation, Social Return on Investment, Sustainable Return on Investment and the new evaluative method of SuHousingImpact.

It has focused on the necessity for an evaluation method which takes into account hidden socio-environmental impacts of housing-led urban regeneration schemes in order to more efficiently be able to gauge the sustainability of a given scheme. It has been suggested that SuROI could be utilised in this context due to its ability to quantify previously intangible socio-environmental benefits.

However, in order to more fully marry up with the UN's SDG11 (Sustainable Development Goal 11), and in order to take into account a continued financial sustainability, this existing method has been refined by tapping into the concept of stakeholder analysis. The qualitative stakeholder analysis has been combined with the more quantitative SuROI. This has created a new method which is able to provide per stakeholder impact values. In addition to this, the concept of time has been introduced via payback period analyses.

Synopsis of what the SuHousingImpact method provides:

1 It enables the gauging of full sustainable impact, from a given outlay of money invested in a housing-led urban regeneration scheme, through an evidence-based proof

DOI: 10.1201/9781003043881-9

The tool uses the foundations of SuROI to show quantitative sustainable impact values. These values are created through a thorough liaison with involved stakeholders; stakeholders who are actively involved and affected by a scheme in question, and who are most likely to be able to thoroughly recount change and impact as part of the involved fieldwork. Also, the usage of relevant indicators and proxies from previously mentioned databases is used to give an overall sustainable value, in quantitative format.

As mentioned, the evidence-based and quantitative approach used within the method means that all aspects pertaining to the triple bottom line, even previously intangible and more difficult to measure ones (Conejos et al., 2013), are able to be measured. The quantification of the impacts results in a ratio which compares the amount of money invested to the resulting impact, both on a per stakeholder basis and for a scheme as a whole.

As stated by Bichard (2016), the notion of an evidence-based proof through the monetisation of the detrimental or the added sustainable value of development schemes "has not featured to date in the literature". Such monetisation reduces complex information into data that can be easily compared and valued (Conejos et al., 2013; Lingane and Olsen, 2004).

The monetisation technique facilitates "the comparison of otherwise incommensurable benefits across different activities" and allows the measurement of "what matters" to end users through a comprehensive method that is both "robust and replicable" (Watson and Whitley, 2016).

2 It can be used to better fulfil sustainability criteria in terms of all three aspects of the triple bottom line and contribute in a more sustainable way to the UN's SDG11

The SuHousingImpact method can better fulfil sustainability criteria which links into the UN's SDG11. It is argued that any regeneration project that fails to evaluate each of the well-established sustainability pillars will not achieve its sustainable development and regeneration objectives (CLG, 2008; Winston, 2009).

The economic climate and its effect on housing associations is a serious challenge to the achievement of the SDGs, since failures in ensuring sufficient support to the housing sector would result in hindering a main pillar of the New Urban Agenda. By actively seeking out and discussing the financial benefits that are coming the way of the not yet involved stakeholder (whose socio-environmental spillovers can be measured through the method), it is possible to prevent schemes being affected by current economic cuts and subsequently ensuring that there are adequate resources to match the ambition of sustainable development (Roberts, 2000), and the UN's SDGs, in particular Goal 11 of the SDGs, by offsetting budgetary cuts through striking agreements with those stakeholders who benefit.

The raising of consensus (and financial support) across as many stakeholders as possible regarding the need for wider interventions relating to a crucial pillar of the SDGs, which is, the allowing of each person the dignity of a decent home, can be carried out by unveiling the larger and long-term impacts on respective budgets. This can be evidenced where each separate stakeholder involved in a housing-led urban regeneration scheme and inputted into the new method as part of the evaluation of a scheme now has the overall monetary calculations available on a per stakeholder basis. This can provide information useful in terms of the offloading of costs onto potential wider stakeholders to ensure a higher rate of financial sustainability, subsequently being capable of turning a cost into a profit. By doing this, the economic problems prevalent at the moment and being felt by many housing associations may be minimised, thus increasing the chances of future financial sustainability. This aspect of the new method could potentially act as a negotiating tool in the quest for more investment into housing-led urban regeneration schemes, thus ensuring financial sustainability going forward.

3 The method fits in well with regard to the era of 'New Public Management'

New Public Management correctly assumes that there are competing agencies where the behaviour of the differing agencies' management is changing. Currently it is the case through New Public Management that each separate stakeholder has its own budget and each separate stakeholder, such as the NHS, is fighting for as much money as it itself can get hold of, and probably looks at its own needs before thinking about needs belonging to other competing stakeholders. As discussed, the notion of sustainability has been actively and widely promoted within sustainable regeneration projects, but apparently with limited sustainability benefits for the different intended beneficiaries involved (Van Bueren and De Jong, 2007).

Therefore, rather than thinking in terms of overall 'public' benefit, it is important to rely on quantifiable methods showing benefits and costs for those different and involved public actors and beneficiaries, exactly what the new method provides. This would potentially be able to show how a specific intervention reflects on each separate stakeholder's budget line. A thorough understanding of the mutual interconnections provided by good quality housing and its related impacts may influence the willingness to fill some of the current gaps in resources for housing provision (Dean et al., 2017a).

The 'SuHousingImpact' tool gauges the quantifiable amount of impact for each stakeholder involved in a housing-led urban regeneration scheme and would show in monetary units whether or not each stakeholder gains or losses from a given project, and additionally, by the cash flow amount involved, thereby marrying up with the concept of 'New Public Management'. Because SuROI does not have this facility or carry out this function, the method is a better fit with regard to the era of 'New Public Management'.

In addition, and in line with the New Public Management concept, if cross co-operation were to be possible amongst stakeholders, this, as promoted as part of the method of this book, could be a particularly useful concept with regard to public policies, theoretically and with specific reference to housing.

4 The stakeholder mapping process used to identify potential wider stakeholders has been shown to be able to be carried out quantitatively, conversely to the usual qualitative manner, in light of the concept of New Public Management

As part of the new 'stage 6' of the method, each individually involved stakeholder is presented with a socio-environmental impact ratio. This ratio shows exactly how much a given stakeholder gains or loses from a scheme in question.

Because of this, the stakeholder mapping process can be shown to have been carried out in a quantitative manner. This directly links into the concept of New Public Management, where, typically, each involved stakeholder relating to a scheme has their own budget to work with. With the separated ratio values, this makes a scheme's impact far more relatable to involved stakeholders, and far more useful when dealing with different budgets, than when compared to an overall scheme value.

In addition to the above, the quantification of the stakeholder mapping process reduces the weaknesses inherently present in such analyses, with the quantitative approach outlined being utilised to reduce the subjectivity of the stakeholder mapping process; the quantified outcomes clearly highlighting the cited "desired outcomes" of Bryson (2004), whilst this book also suggests that because the interest and influence of stakeholders and how much a particular stakeholder stands to gain or lose from a given project is quantified, Reed et al.'s (2009) point regarding the need to show extendable matrices that show interest and influence of stakeholders is fulfilled.

5 The method introduces the concept of time, which leads to the new method being used as a strategic, management- or governance-based tool, whilst additionally ensuring continued economic sustainability

The new method is not only an evaluative tool, but can be used as a strategic management or governance tool (also see Dean et al., 2017a). It does this through being able to use the concept of time, via payback period analysis, to gauge whether to invest and indeed what type(s) of investment to carry out. Also, this is a very useful concept for strategic decision making, management or governance. Using these calculations, it would be able to prioritise schemes based on how quickly the payback periods come along or alternatively it would be able to potentially cancel plans for any schemes which take too long to break even and it would additionally be able to be used to target certain groups of stakeholders that housing associations or housing-led urban regeneration facilitators wish to target as part of a scheme; either such stakeholders who previously have not benefitted from any such schemes or stakeholders who are currently benefitting from schemes at a cost to the housing association and who could potentially supplement such costs, again thus ensuring continued economic sustainability.

6 It can be used to ensure less wasting of money on schemes and can be used to plug the gap created by the recently experienced economic shortfall which is impacting on housing associations, tenants and communities alike

Previously, there has been a "lack of effective tools and robust methodologies capable of capturing the full social value generated by the housing association sector" (Fujiwara and HACT, 2013). Usage of the SuHousingImpact method ensures less wasting of money and fewer poor decisions being carried out, through its provision of a full sustainable evaluation of socio-environmental impact.

The evidence-based and quantitative approach used within the method means that all aspects, even previously intangible and more difficult to measure ones (Conejos et al., 2013), are now able to be measured. This translates to there being the maximum amount of information available on social and environmental benefits "of central concern to individuals and communities" (Vardakoulias, 2013). With this amount of information now being available to decision makers in the housing sector field, this, in turn, means that it is more difficult, theoretically, to waste money through poor investments. This being contrary to previous decision making processes, where money has been spent on areas which then does not in turn make any difference to an area socially or environmentally.

The tool additionally plugs the gap created by the recently created economic shortfall by involving more financial contributions from actual and/or potential stakeholders. Because the tool actively identifies the full range of stakeholders benefitting from a scheme in question and quantifies the respective costs and benefits pertaining to each stakeholder, in a way that previous tools have failed to carry out, stakeholders have an evidence-based proof of how they benefit. Such an evidence-based approach being presented to stakeholders showing clearly what is received by way of benefits is a much more reliable way of enabling an understanding of impact. Through the evidence-based method, stakeholders are able to realise that positive impacts resulting from a particular scheme can far outweigh the cost of a potential financial contribution towards the economic sustainability of the said scheme, thus enabling all parties to continue to benefit. In a particular scheme's absence, costs incurred may outweigh the previous benefits, making it economically worse off for both the scheme facilitator and stakeholder(s) in question. This aspect of SuHousingImpact can ensure that Roberts's (2000) point regarding the necessity for adequate resources to match the ambition of sustainable development is covered.

7 The tool is able to be used to evaluate historical housing-led urban regeneration schemes, but can also be used in a predictive way, on future schemes, or indeed could be used on any stakeholder-led scheme within any domain

The SuHousingImpact method can evaluate in both an ex post and an ex ante manner.

Historical schemes from the past can be assessed. However, the method can also be used in a predictive way, assessing potential impacts of schemes yet to be unveiled. This can be carried out because of the amount of valuation data the method taps into. Indeed, the guide for SROI, on which the tool is heavily based, states that SROI can be carried out ex ante, as a forecasting or predictive method (Nicholls et al., 2012). The tool is useful in terms of predicting or defining what is wanted from potential future contracts, whilst stakeholder data collected infield can be used to not only evaluate social and environmental change caused by previously completed projects, but also to predict future change from planned projects. As noted earlier, it is also prudent to again mention that the method focuses on the potential to look in far greater depth at future economic sustainability than has previously been covered within SuROI. Additionally, the tool has significant potential for future research because as the tool is stakeholder focused and stakeholder led, it can be applied to any stakeholder-related scheme. This includes other subject areas and domains such as "health care or sport facilities, by exploring a variety of different stakeholders and respective benefits and burdens and testing their willingness to be subsequently engaged in the production of public goods" (Dean et al., 2017b).

8 Through the method's aforementioned use as a strategic decision making or management tool, schemes can be planned in, prioritised or carried out in a targeted and strategic manner.

> As has been highlighted for SROI, on which the method is based, the monetised indicators can help management analyse what might happen if they change their strategy, as well as allowing them to evaluate the suitability of that strategy in generating social [or environmental] returns, or whether there may be better means of using their resources. (Maughan, 2012).

However, in addition to the above, through the new stage 6 of the method, the various payback period calculations ensure that SuHousingImpact can be used as part of a strategic decision making, management or governance tool.

To this end, it would be able to plan in or prioritise schemes based on how quickly respective payback periods come along or alternatively it would be possible to potentially cancel plans for any schemes which take too long to break even.

The tool would additionally be able to be used to target certain groups of stakeholders that housing associations or housing-led urban regeneration facilitators wish to target as part of a scheme through the per stakeholder impact ratios; either such stakeholders who previously have not benefitted from any such schemes or stakeholders who are currently benefitting from schemes at a cost to the housing association and who could potentially supplement such costs, thus ensuring continued economic sustainability.

The tool could be used to make logical decisions on where to start work, and in what order, based on where has the greatest impact, or a specific scheme could be chosen to be unveiled which is tailored to certain aspects of the social or environmental spectrum, in order to enable a specific community, previously suffering from a particular social or environmental problem, to have a greater impact in that specific area.

Potential future works could be prioritised depending on what kinds of impact the housing-led urban regeneration facilitator/organisation in question wants to see within a designated area. As mentioned within the interviews, it is also possible to work backwards from the kinds of impact wanted, towards what scheme factors would produce such an impact. This could lead to works specifications being sustainability focussed.

9 It can be used for modelling purposes, for publicity purposes and can also be used alongside existing tools

The tool can be used to model different scenarios, for example, showing various different plans, eventualities or scheme impacts – this acting as a way to model and compare results of different scheme ideas.

In addition, data could be inputted into the tool to be used to predict varying scenarios. Through this, the impacts potentially being generated by a potential future scheme being put up before, for example, a local authority, regulatory body or potential funder, can be shown as part of the publicity process for such a scheme, in order to promote, advertise or sell the scheme in question to the local authority, regulatory body or potential funder in question.

Through the different stakeholder values generated within the new stage 6 of the method, the sustainable value that, for example, a local authority were to get back from a potential scheme idea is able to be generated. Also if donations to schemes were being sought, payback periods could be used to convince funders to fund.

The SuHousingImpact tool can also be used alongside existing tools as a further proof of impacts, useful to be shown as part of any publicity or funding seeking process. Despite there having been many evaluation methods utilised within the field of housing-led urban regeneration schemes, there remained a gap in the knowledge base; this gap being namely that no tailored tool existed to quantitatively measure the hidden or previously intangible social and environmental benefits that typically result from a housing-led urban regeneration scheme. In addition, there would appear to be the need to re-orient existing assessment methods towards a more strategic and long-term sustainability suitable to engage potential stakeholders to provide financial sustainability going forward.

This book now provides, for public consumption, a tool which can make visible previously intangible and previously invisible socio-environmental spillovers that typically result from housing-led urban regeneration schemes.

However, in addition, the new method created, the SuHousingImpact tool, also ensures continued financial sustainability going forward, particularly useful within the current economic climate. In addition, the introduction of the concept of time into the equation also enables prospective users to be able to not only fully evaluate schemes, but in addition to be able to use the said tool as a strategic management or governance tool.

It is the authors' hope that this tool enables housing-led urban regeneration to ultimately become more sustainable and for the built environment to become something of which we can all be proud of. Something of which is so important to every single one of us.

References

Adams, W. M. (2006). *The Future of Sustainability. Re-thinking Environment and Development in the Twenty-First Century*. The World Conservation Union. Available at: http://cmsdata.iucn.org/downloads/iucn_future_of_sustanability.pdf (Accessed 01/02/2018).

Adger, N. and Winkles, A. (2007). 'Vulnerability, poverty and sustaining wellbeing', in *Handbook of Sustainable Development*. Cheltenham: Edward Elgar Publishing Limited.

Allen, W. and Kilvington, M. (2002). 'Learning and working together for the environment: Applying the integrated systems for knowledge management approach', *Development Bulletin*, 58, pp. 106–111.

Anderson, S. R., Bryson, J. M. and Crosby, B. C. (1999). *Leadership for the Common Good Fieldbook*. St. Paul: University of Minnesota Extension Service.

Andrews, C. J. (2007). 'Rationality in policy decision making', in F. Fischer, Gerald J. Miller and Mara S. Sidney (Eds) *Handbook of Public Policy Analysis: Theory, Politics and Methods*. Boca Raton, FL: CRC Press, pp. 43–62.

Ang, S. L. and Wilkinson, S. J. (2008). 'Is the social agenda driving sustainable property development in Melbourne, Australia?' *Property Management*, 26(5), pp. 331–343.

Arli, D. and Zappala, G. (2009). *Why Do Companies Ignore Measuring the Social Impact of Their Corporate Community Involvement Programs?* Available at: www.csi.edu.au (Accessed 07/09/2017).

Arvidson, M., Lyon, F., McKay, S. and Moro, D. (2013). 'Valuing the social? The nature and controversies of measuring social return on investment', *Voluntary Sector Review*, 4(1), pp. 3–18.

Ashford, N. A. and Hall, R. P. (2011). *Technology, Globalization, and Sustainable Development: Transforming the Industrial State*. New Haven, CT: Yale University Press.

Aspden, P., Ball, A., Roberts, M. and Whitley, T. (2012). 'A holistic evidence-based approach to retrofit in Social Housing'. Paper presented at the Retrofit 2012 Conference 24th–26th January. Salford: University of Salford.

Atkinson, G. (2008). 'Sustainability, the capital approach and the built environment', *Building Research & Information*, 36(3), pp. 241–247.

Atkinson, R. and Moon, G. (1994). *Urban Policy in Britain: The City, the State and the Market*. London: MacMillan.

Audit Commission (2003). *Economic and Community Regeneration: Learning from Inspection*. London: Audit Commission National Study.

Audit Commission Publication (2009). *Building Better Lives*. Available at: http://www.auditcommission.gov.uk/nationalstudies/localgov/buildingbetterlives/Pages/casestudies.aspx (Accessed 10/11/2017).

Australian Department of Sustainable Environment (2007). *Stakeholder Analysis*. Available at: www.dse.vic.gov.au (Accessed 09/02/2018).

Awomelwe, A. and Ogundele, O. (2008). *The Importance of the Payback Method in Capital Budgeting Decision*. Unpublished MBA thesis. Sweden: Blekinge Institute of Technology.

Babu, S. and Pinstrup-Andersen, P. (2009). 'Social innovation and entrepreneurship: Developing capacity to reduce poverty and hunger', in J. von Braun, R. Vargas Hill & R. Pandya-Lorch (Eds) *The Poorest and Hungry: Assessments, Analyses, and Actions. An IFPRI 2020 Book*. Washington, DC: International Food Policy Research Institute, pp. 541–548.

Bäckstrand, K. (2005). 'Accountability and legitimacy of networked governance. Public-Private partnerships for sustainable development'. *Paper presented at the 2005 Berlin Conference "International Organizations and Global Environmental Governance"*, 2–3 December 2005. Sweden: Lund University.

Bailey, N. (2010). 'Understanding community empowerment in urban regeneration and planning in England: Putting policy and practice in context', *Planning Practice and Research*, 25(3), pp. 317–332.

Ballinger, S. (2002). *Home Sick: Shelter and Bradford and Bingley's Campaign for Healthy Homes*. London: Shelter.

Barnekov, T., Boyle, R. and Rich, D. (1989). *Privatism and Urban Policy in Britain and the United States*. Oxford: Oxford University Press.

Barzelay, M. (2001). *The New Public Management: Improving Research and Policy Dialogue*. Berkeley & New York: University of California Press & Russell Sage Foundation.

Basham, M., Shaw, S., Barton, A. and Torbay, H. (2004). *Central Heating: Uncovering the Impact on Social Relationships in Household Management*. Exeter: Peninsula Medical School, Eaga Partnership Charitable Trust.

Battle, G. (2014). *Doing Good Is Good Business Roundtable: Public Services (Social Value) Act – Review and Cross Party Recommendations*. London: Sustainable Business Partnership.

Bell, S. and Morse, S. (2003). *Measuring Sustainability. Learning from Doing*. London: Earthscan.

Bennett, J. and Crudgington, A. (2003). 'Sustainable development: Recent thinking and practice in the UK', *Engineering Sustainability*, 156(Issue ESI), pp. 27–32.

Berry, J., McGreal, S. and Deddis, B. (Eds). (1993). *Urban Regeneration: Property Investment and Development*. London: E. and F. N. Spon.

Berger, I. A., Cunningham, P. H. and Drumwright, M. E. (2004). 'Social alliances: Company/non-profit collaboration', *California Management Review*, 47, pp. 58–90.

Bhatti, M. (1999). 'The meaning of gardens in an age of risk', in T. Chapman and J. Hockey (Eds) *Ideal Homes? Social Change and Domestic Life*. London: Routledge, pp. 181–194.

Bichard, E. (2015). *Developing an Approach to Sustainable Return on Investment in the UK, Brazil and the USA*. A report for the Royal Institution of Chartered Surveyors Research Trust. London: RICS Research Trust.

Bichard, E. (2016). 'Sustainable return on investment: Towards a method of valuing social and environmental change in the built environment', *Proceedings of the 2nd Annual International Conference on Social Sciences*, 2, pp. 866–874.

Bingham, G., Bishop, R., Brody, M., Bromley, D., Clark, E., Cooper, W., Costanza, R., Hale, T., Hayden, G., Kellert, S., Norgaard, R., Norton, B., Payne, J., Russell, C. and Suter, G. (1995). 'Issues in ecosystem valuation: Improving information for decision making', *Ecological Economics*, 14, pp. 73–90.

Blatt, J. M. (1979). 'Investment evaluation under uncertainty', *Financial Management*, 8(2), pp. 66–81.

Blessing, A. (2015). 'Public, private, or in-between? The legitimacy of social enterprises in the housing market', *VOLUNTAS: International Journal of Voluntary and Nonprofit Organizations*, 26(1), pp. 198–221.

Bogard, C. J., McConnell, J. J., Gerstel, N. and Schwartz, M. (1999). 'Homeless mothers and depression: Misdirected policy', *Journal of Health and Social Behaviour*, 40(1), pp. 46–62.

Bond, L., Kearns, A., Mason, P., Tannahill, C., Egan, M. and Whitley, E. (2012). 'Exploring the relationships between housing, neighbourhoods and mental wellbeing for residents of deprived areas', *BMC Public Health*, 12, p. 48.

Bostani, M. and Malekpoor, A. (2012). 'Critical analysis of Kaldor-Hicks efficiency criterion, with respect to moral values, social policy making and incoherence', *Advances in Environmental Biology*, 6(7), pp. 2032–2038.

Bourne, L. and Walker, D. H. T. (2006). 'Visualizing stakeholder influence – Two Australian examples', *Project Management Journal*, 37(1), pp. 5–22.

Bowey, R. (1997). *Literature Review on Urban Renewal*. Adelaide: South Australian Department of Human Services.

Bowling, A. (1997). *Measuring Health: A Review of Quality of Life Measurement Scales*. Milton Keynes: Open University Press.

Boyko, C. T., Gaterell, M. R., Barber, A. R. G., Brown, J., Bryson, J. R., Butler, D., et al. (2012). 'Benchmarking sustainability in cities: The role of indicators and future scenarios', *Global Environmental Change: Human and Policy Dimensions*, 22(1), pp. 245–254.

Brammer, S. and Millington, A. (2004). 'The development of corporate charitable contributions in the UK: A stakeholder analysis', *Journal of Management Studies*, 41(8), pp. 1411–1434.

Brandon, P. S. and Lombardi, P. (2011). *Evaluating Sustainable Development in the Built Environment*. Oxford: Wiley-Blackwell.

Brennan, A., Rhodes, J. and Tyler, P. (1997). *Evaluation of the Single Regeneration Budget Challenge Fund*. London: DETR.

Brennan, A., Rhodes, J. and Tyler, P. (1998). *Evaluation of the Single Regeneration Challenge Fund Budget: A Partnership for Regeneration: An Interim Evaluation*. London: The Stationery Office.

Brennan, A., Rhodes, J. and Tyler, P. (1999). 'The distribution of SRB challenge fund expenditure in relation to local-area need in England', *Urban Studies*, 36(2), pp. 2069–2084.

Bridgeman, J., Maple, P., Murdock, A., Hardy, S. and Townley, C. (2016). 'Demonstrating the social value of a schools engagement programme: Introducing young people to the construction professions', in P. W. Chan and C. Neilson (Eds) *Proceedings 32nd Annual ARCOM Conference*, Manchester, Association of Researchers in Construction Management, 5–7 September 2016, pp. 1007–1016.

Bridgeman, J., Murdock, A., Maple, P., Townley, C. and Graham, J. (2015). Putting a value on young people's journey into construction: Introducing SROI at Construction Youth Trust', in A. Raiden and E. Aboagye-Nimo (Eds) *Proceedings 31st Annual ARCOM Conference, Association of Researchers in Construction Management*, Lincoln, 7–9 September, pp. 207–216.

Brown, C. (2006). *Building*. Available at: http://www.building.co.uk/story.asp?storycode=3062794 (Accessed 01/10/17).

Brownill, S. (1990). *Developing London's Docklands: Another Great Planning Disaster?* London: Paul Chapman Publishing Ltd.

Brugha, R. and Varvasovszky, Z. (2000). 'Stakeholder analysis: A review', *Health Policy and Planning*, 15(3), pp. 239–246.

Bryson, J. (1995). *Strategic Planning for Public and Non-Profit Organization* (Rev. Ed). San Francisco, CA: Jossey-Bass Publishers.

Bryson, J. M. (2004). 'What to do when stakeholders matter: Stakeholder identification and analysis techniques', *Public Management Review*, 6(1), pp. 21–53.

Burke, C. and King, A. (2015). 'Generating social value through public sector construction procurement: A study of local authorities and SMEs', in A. Raiden and E. Aboagye-Nimo (Eds) *Proceedings 31st Annual ARCOM Conference*, 7–9 September 2015. Lincoln: Association of Researchers in Construction Management, pp. 387–396.

Caldwell, J., McGowan, S., McPhail, J., McRae, C., Morris, G. and Murray, K. (2001). *Glasgow Warm Homes Study: Final Report*. Glasgow: Glasgow City Council Housing Services.

Caldwell, L. K. and Weiland, P. S. (1996). *International Environmental Policy: From the Twentieth to the Twenty-first Century*. (3rd Ed). Durham, NC: Duke University Press.

Cambridge Economic Associates (CEA). (2001). *Department for Social Development People and Place a Strategy for Neighbourhood Renewal Working Paper 2 Urban Regeneration in Northern Ireland –Policy Measurement*. Cambridge: CEA.

Carley, M. and Kirk, K. (1998). *Sustainable by 2020? A Strategic Approach to Regeneration in Britain's Cities*. Bristol: Policy Press.

Carter, C. R. and Rogers, D. S. (2008). 'A framework of sustainable supply chain management: Moving toward new theory', *International Journal of Physical Distribution & Logistics Management*, 38(5), pp. 360–387.

Carter, K. and Fortune, C. (2007). 'Sustainable development policy perceptions and practice in the UK social housing sector', *Construction Management and Economics*, 25, pp. 399–408.

Carter, T. (2012). 'Neighbourhood improvement: The role of housing and housing institutions', in *The International Encyclopaedia of Housing and Home*. Oxford: Elsevier, pp. 67–72.

Centre for Local Economic Strategies (CLES). (2015). *Progressive Housing Associations*. Available at: http://www.cles.org.uk/wp-content/uploads/2014/04/CLES-10-Progressive-Housing-Associations.pdf (Accessed 07/11/2017).

Chang, A., Chih, Y.-Y., Chew, E. and Pisarski, A. (2013). 'Reconceptualising mega project success in Australian Defence: Recognising the importance of value co-creation', *International Journal of Project Management*, 31(8), pp. 1139–1153.

Chartered Institute of Housing. (2013). *Regenerating Our Communities: A Snapshot in Time*. Edinburgh: CIH Scotland.

Chen, Y., Song, Y., Bowker, S. and Hamilton, A. (2012). 'The SURegen workbench: A web-based collaborative regeneration tool', *International Journal of E-Planning Research (IJEPR)*, 1(2), pp. 44–64.

Chevin, D. (2013). *Social Hearted*. Available at: http://www.d4multimedia.com/genesis/Social-hearted.pdf (Accessed 01/02/2018).

Chevin, D. (2014). *Gearing up for Giving Something Back*. Available at: http://www.constructionmanagermagazine.com/agenda/social-value-gearing-giving-back (Accessed 31/07/2017).

Chipman, J. S. (1987). 'Compensation principle', in *The New Palgrave Dictionary of Economics*, (2nd Ed). London: Palgrave MacMillan.

Church, C. and Young, S. (2001). 'The United Kingdom: Mainstreaming, mutating or expiring?' in W. M. Lafferty (Ed) *Sustainable Communities in Europe*. London: Earthscan Publications Ltd.

Clapham, D. (2014). *Regeneration and Poverty in Wales: Evidence and Policy Review*. Sheffield: Centre for Regional Economic and Social Research.

Clapham, D., Walker, R., Meen, G., Thake, S. and Wilcox, S. (1996). *Building Homes Building Jobs: Housing and Economic Renewal.* London: National Housing Forum.

Clark, W. and Munn, T. (Eds). (1986). *Sustainable Development of the Biosphere.* Cambridge: Cambridge University Press.

Clements, A. (Ed). (2005). *Game Plan: Housing, Planning and the Economy.* Cardiff: Chartered Institute of Housing, Cymru.

Colantonio, A. and Dixon, T. (2009). *Measuring Socially Sustainable Urban Regeneration in Europe.* Oxford: Oxford Institute for Sustainable Development (OISD), Oxford Brookes University.

Colantonio, A. and Dixon, T. (2010). *Urban Regeneration and Social Sustainability: Best Practice from European Cities.* Oxford: Wiley.

Communities and Local Government (CLG). (2008). *Transforming Places; Changing Lives: A Framework for Regeneration.* London: CLG.

Communities and Local Government (CLG). (2009). *Transforming Places; Changing Lives – A Framework for Regeneration: Summary of Consultation Responses.* London: CLG.

Conejos, S. Langston, C. and Smith, J. (2013). 'AdaptSTAR Model: A climate – Friendly strategy to promote built environment sustainability', *Habitat International,* 37, pp. 95–103.

Costanza, R., d'Arge, R., de Groot, R.S., Farber, S., Grasso, M., Hannon, B., Limburg, K., Naeem, S., O'Neill, R.V., Paruelo, J., Raskin, R.G., Sutton, P. and van den Belt, M. (1997). 'The value of the world's ecosystem services and natural capital', *Nature,* 387, pp. 253–260.

Couch, C. (1990). *Urban Renewal Theory and Practice.* Basingstoke: Macmillan.

Couch, C. and Dennemann, A. (2000). 'Urban regeneration and sustainable development in Britain', *Cities,* 17, pp. 137–147.

Couch, C., Sykes, O. and Boerstinghaus, W. (2011). 'Thirty years of urban regeneration in Britain, Germany and France: The importance of context and path dependency', *Progress in Planning,* 75(1), pp. 1–52.

Daily, G. C. (Ed). (1997). *Nature's Services: Societal Dependence on Natural Ecosystems.* Washington, DC: Island Press.

Daily, G. C., Soderquist, T., Aniyar, S., Arrow, K., Dasgupta, P., Ehrlich, P. R., Folke, C., Jannson, A., Jansson, B-O., Kautsky, N., Levin, S., Lubchenco, J., Maler, K-G., David, S., Starrett, D., Tilman, D. and Walker, B. (2000). 'The value of nature and the nature of value', *Science,* 289, pp. 395–396.

Davidson, M. (2009). 'Social sustainability: A potential for politics?' *Local Environment,* 14, pp. 607–619.

Dayananda, D., Irons, R., Harrison, S., Herbohn, J. and Rowland, P. (2002). *Capital Budgeting. Financial Appraisal of Investment Projects.* Cambridge: Cambridge University Press.

Deakin, N. and Edwards, J. (1993). *The Enterprise Culture and the Inner City.* London: Routledge.

Deakin, M. (2009). 'A Community-Based Approach to Sustainable Urban Regeneration', *Journal of Urban Technology,* 16(1), pp. 91–112. DOI: 10.1080/10630730903090354.

Dean, K., Trillo, C. and Bichard, E. (2017a). 'Assessing the value of housing schemes through sustainable return on investment: A path towards sustainability-led evaluations?', *Sustainability,* 9(12), p. 2264. Available at: http://dx.doi.org/10.3390/su9122264 (Accessed 01/03/2018).

Dean, K., Trillo, C. and Bichard, E. (2017b). 'Housing and urban regeneration: Honing an existing assessment method through the design science methodology', in C. Pathirage,

U. Kulatunga, Y. Ji, R. N. Gameson, C. E. Udeaja, C. Trillo, M. Takhtravanchi and B. Allali (Eds) *13th International Postgraduate Research Conference 2017*, Salford: University of Salford, 14–15 September 2017.

Dean, K. (2019). *Towards a Tool to Gauge the Success Levels of Housing-Led Urban Regeneration Schemes*. PhD thesis, University of Salford. Salford.

Deepak, M. (2011). *Social Housing as a Multi-Stakeholder Partnership Approach*. Unpublished MPA thesis. FHR: Paramaribo.

De Groot, R. (1992). *Functions of Nature: Evaluation of Nature in Environmental Planning, Management and Decision Making*. Groningen: Wolters-Noordhoff.

De Groot, R. (1994). 'Environmental functions and the economic value of natural ecosystems', in Jansson et al. (Eds) *Investing in Natural Capital: The Ecological Economics Approach to Sustainability*. Washington, DC: Island Press, pp. 151–168.

De Groot, R., van der Perk, J., Chiesura, A. and Marguliew, S. (2000). 'Ecological functions and socio-economic values of critical natural capital as a measure for ecological integrity and environmental health', in P. Crabbe, A. Holland, L. Ryszkowski and L. Westra, L. (Eds) *Implementing Ecological Integrity: Restoring Regional and Global Environmental and Human Health*. NATO-Science Series, IV. Earth and Environmental Sciences, vol. 1. Dordrecht/Boston, MA/London: Kluwer Academic Publishers, pp. 191–214.

De Groot, R., Wilson, M. and Boumans, R. (2002). 'A typology for the classification, description and valuation of ecosystem functions, goods and services', *Ecological Economics*, 41, pp. 393–408.

Department for Communities and Local Government (DCLG). (2008). *Transforming Places; Changing Lives: A Framework for Regeneration*. London: DCLG.

Department for Communities and Local Government (DCLG). (2011). *Regeneration to Enable Growth: What the Government Is Doing in Support of Community Led-Regeneration*. London: DCLG.

Department for Communities and Local Government. (2014). *The Importance of Good Design*. http://planningguidance.communities.gov.uk/blog/guidance/design/the-importance-of-good- design/.

Department for Environment, Food and Rural Affairs (DEFRA). (2007). *An Introductory Guide to Valuing Ecosystem Services*, Department for Environment, Food and Rural Affairs. London: DEFRA.

Department for Environment, Food and Rural Affairs (DEFRA). (2011). *Encouraging Businesses to Manage Their Impact on the Environment. Business and Enterprise and Environment*. London: DEFR.

Department of Health. (1998). *Our Healthier Nation A Contract for Health*. Available at: https://www.gov.uk/government/uploads/system/uploads/attachment_data/file/265721/title.pdf (Accessed 01/11/2017).

Department of the Environment (DoE). (1977). *Policy for the Inner Cities*. London: HMSO.

Department of the Environment (DoE). (1997). *The SRB Evaluation Framework*; Discussion Article 83; Department of Land Economy, Cambridge: University of Cambridge.

Department of the Environment, Transport and the Regions (DETR). (1998). *Sustainable Regeneration Good Practice Guide*. London: HMSO.

Department of the Environment, Transport and the Regions (DETR). (1999). *Building a Better Quality of Life: A Strategy for Sustainable Development for the UK*. London: The Stationery Office.

Department of the Environment, Transport and the Regions (DETR). (2000). *Towards a Language of Sustainable Development*. London: The Stationery Office.

Dernbach, J. C. (1998). 'Sustainable development as a framework for national governance', *Case Western Reserve Law Review*, 49, pp. 1–103.

Diamond, J., Liddle, J., Southern, A. and Townsend, A. (2006). *Managing the City*. London: Routledge.

Diesendorf, M. (2000). 'Sustainability: The corporate challenge of the 21st century', *Sustainability and Susainable Development*, 2, pp. 19–37.

Dixon, T., Otsuka, N. and Abe, H. (2013). 'Critical Success Factors in urban brownfield regeneration: bringing hardcore sites in Manchester and Osaka back into use', in M. Leary and J. McCarthy (Eds) *The Routledge Companion to Urban Regeneration*. London: Routledge, pp. 241–250.

Dowler, C. (2009). *Shared ownership sales slump*. Inside Housing (28 August). Available at: http://www.insidehousing.co.uk/shared-ownership-sales-slump/6506100.article (Accessed 01/02/2018).

Drakakis-Smith, D. (1995). 'Third world cities: Sustainable urban development, 1', *Urban Studies*, 32(4–5), pp. 659–677.

Du Pisani, J. A. (2006). 'Sustainable development – Historical roots of the concept', *Environmental Sciences*, 3(2), pp. 83–96.

Du Plessis, C. (2005). 'Action for sustainability: Preparing an African plan for sustainable building and construction', *Building Research & Information*, 33(5), pp. 405–415.

Durie, S., Hutton, E. and Robbie, K. (2007). *Investing in Impact: Developing Social Return on Investment*. Available at: http://www.socialimpactscotland.org.uk/media/1200/SROI-%20investing%20in%20impact.pdf (Accessed 10/12/2017).

Ebrahim, A. and Rangan, V. K. (2010). *The Limits of Nonprofit Impact: A Contingency Framework for Measuring Social Performance*. Boston, MA: Harvard Business School.

Eden, C. and Ackermann, F. (1998). *Making Strategy: The Journey of Strategic Management*. London: Sage Publications.

Edwards, B. and Turrent, D. (Eds) (2000). *Sustainable Housing – Principles & Practice*. London: E & F N Spon.

Egan, M., Katikireddi, S., Kearns, A., Tannahill, C., Kalacs, M. and Bond, L. (2013). 'Health effects of neighborhood demolition and housing improvement: A prospective controlled study of 2 natural experiments in urban renewal', *American Journal of Public Health*, 103(6), pp. e47–e53.

Ehrlich, P. R. (1968). *The Population Bomb*. New York: Ballantine Books.

Ehrlich, P. R. and Ehrlich, A. (1992). 'The value of biodiversity', *Ambio*, 21(3), pp. 219–226.

Ekins, P., Dresner, S. and Dahlström, K. (2008). 'The four-capital method of sustainable development evaluation', *European Environment*, 18(2), pp. 63–80.

Ekins, P., Simon, S., Deutsch, L., Folke, C. and De Groot, R. (2003). 'A framework for the practical application of the concepts of critical natural capital and strong sustainability', *Ecological Economics*, 44, pp. 165–185.

Elias, A. A., Cavana, R. Y. and Jackson, L. S. (2002). 'Stakeholder analysis for R&D project management', *R&D Management*, 34(2), pp. 301–310.

Elkington, J., Tickell, S. and Lee, M. (2007). *SustainAbility. 20 Years of Global Leadership*. London: SustainAbility.

Emerson, J. and Twersky, F. (1996). *New Social Entrepreneurs: The Success, Challenge and Lessons of Non-profit Enterprise Creation*. San Francisco, CA: Roberts Foundation, Homeless Economic Development Fund.

Emerson, J., Wachowicz, J. and Chun, S. (2001). *Social Return on Investment: Exploring Aspects of Value Creation in the Nonprofit Sector*. Cambridge: Harvard Business School. Available at: http://hbswk.hbs.edu/archive/1957.html (Accessed 16/08/2017).

Emmelin, L. (1972). 'The Stockholm conferences', *Ambio*, 1(4), pp. 135–140.

Energy Environment and Sustainable Development within the Fifth Framework Programme of the European Union. (2006). LUDA E-compendium: Handbook E3, *A Community-Based Approach to Sustainable Urban Regeneration: The Key Role of Participation and Futures Workshops from City of Tomorrow & Cultural Heritage from the Programme Energy, Environment and Sustainable Development within the Fifth Framework Programme of the European Union*. http://www.luda-project.net/compendium/pdf/hbe3_060206.pdf.

Entec. (2004). *Study into the Environmental Impacts of Increasing the Supply of Housing in the UK. Final Report*. London: DEFRA.

Environmental Audit Committee. (2005). *Housing: Building a Sustainable Future*. Volume 1. London: House of Commons, Stationery Office.

Ercan, M. A. (2011). 'Challenges and conflicts in achieving sustainable communities in historic neighbourhoods of Istanbul', *Habitat International*, 35(2), pp. 295–306.

Eskerod, P. and Jepsen, A. L. (2013). *Project Stakeholder Management*. London: Gower.

European Central Bank. (2003). *Structural Factors in the EU Housing Markets*. Frankfurt: ECB.

European Commission. (2002). *The World Summit on Sustainable Development. People, Planet, Prosperity*. Luxembourg: Office for Official Publications of the European Communities.

Evans, B., Elesei, P., Rosenfeld, O., Roll, Gulnara, F. and Keiner, M. (2016). 'Habitat III – Towards a New Urban Agenda', *The Planning Review*, 52(1), pp. 86–91.

Evans, G. L. and Shaw, P. (2001a). *Study into the Social Impact of Lottery Good Cause Spending*. London: Department for Culture Media and Sport.

Evans, G. L. and Shaw, S. (2001b). 'Urban leisure and transport: Regeneration effects', *Journal of Leisure Property*, 1(4), pp. 350–372.

Evans, J. (2012). 'Sustainable regeneration', in *International Encyclopaedia of Housing and Home*. London: Elsevier, pp. 124–128.

Evers, A., Ewert, B. and Brandsen, T. (2014). *Transnational Patterns and Approaches from 20 European Cities*. Liege: EMES European Research Network.

Fischer, F., Miller, G. and Sidney, M. (Eds) (2007). *Handbook of Public Policy Analysis: Theory, Politics and Methods*. Boca Raton, FL: CRC Press.

Folke, C. (2006). 'Resilience: The emergence of a perspective for social-ecological systems analyses', *Global Environmental Change*, 16, pp. 253–267.

Forum for the Future. (2017). *Five Capitals*. Available at: https://www.forumforthefuture.org/sites/default/files/project/downloads/five-capitals-model.pdf (Accessed 04/01/2018).

Freeman, H. E. (1977). 'The present status of evaluation research', in M. A. Guttentag and S. Saar (Eds) *Evaluation Studies Review Annual*, vol. 2 Beverly Hills, CA: Sage Publications, pp. 17–51.

Freeman, R. E. (1984). *Strategic Management: A Stakeholder Approach*. Marshfield, MA: Pitman.

Freeman, R. E. (1999). 'Divergent stakeholder theory', *Academy of Management Review*, 24(2), pp. 233–236.

French, S. (1988). *Decision Theory – An Introduction to the Mathematics of Rationality*. Chichester: Ellis Horwood Limited.

Fujiwara, D. and the Housing Associations' Charitable Trust (HACT). (2013). *The Social Impact of Housing Providers*. London: HACT.

Gibson, M. and Kocabas, A. (2001). *London: Sustainable Regeneration- Challenge and Response*. Paper presented to the Rendez-vous Istanbul. 1. International Urban Design Meeting. Istanbul: Mimar Sinan University.

Glennerster, H. and Turner, T. (1993). *Estate Based Housing Management: An Evaluation*. London: Her Majesty's Stationery Office.

Global Knowledge Partnership (GKP). (2003). *Multi Stakeholder Partnerships Issue Paper 2003*. Kuala Lumpur: Global Knowledge Partnership.

Global Value Exchange (GVE). (2017). *Global Value Exchange Database*. Available at: http://www.globalvaluexchange.org (Accessed 01/02/2018).

Gregory, D. and Martin, S. (1988). 'Issues in the evaluation of inner city programmes', *Local Econ*, 2(4), pp. 237–249.

Grimble, R. and Chan, M. K. (1995). 'Stakeholder analysis for natural resource management in developing countries; some practical guidelines for making management more participatory and effective', *Natural Resource Foru*, 19(2), pp. 113–124.

Guba, E. G. and Lincoln, Y. S. (1981). *Effective Evaluation: Improving the Usefulness of Evaluation Results through Responsive and Naturalistic Approaches*. San Francisco, CA: Jossey-Bass.

Gupta, A. (1995). 'A stakeholder analysis approach for interorganizational systems', *Industrial Management and Data Systems*, 95(6), pp. 3–7.

Ha, S-K. (2007). 'Housing regeneration and building sustainable low income communities in Korea', *Habitat International*, 31(1), pp. 116–129.

Hajer, M. A. and Wagenaar, H. (Eds). (2003). *Deliberative Policy Analysis. Understanding Governance in the Network Society*. Cambridge: Cambridge University Press.

Hall Aitken. (2011). 'Evaluation of MSV & SROI Project final report', Glasgow, July 2011. Available at: http://www.scotland.gov.uk/Resource/Doc/48453/0124192.pdf (Accessed 20/10/2017).

Hall, R. P. and Ashford, N. A. (2012). *Primer on the Emergence and Evolution of Sustainable Development (1951 to 2012)*. Virginia Tech, VA. Available at: https://ralphphall.files.wordpress.com/2012/11/primer_sustdev_2012-11-26.pdf (Accessed 01/12/2017).

Hambleton, R. and Thomas, H. (Eds). (1995). *Urban Policy Evaluation: Challenge and Change*. London: Paul Chapman Publishing.

Haran, M., Newell, G., Adair, A., McGreal, S. and Berry, J. (2011). 'The performance of UK regeneration property within a mixed asset portfolio', *Journal of Property Research*, 28(1), pp. 75–95.

Hardin, G. (1968). 'The tragedy of the commons', *Science*, 162(3859), pp. 1243–1248.

Harding, A. and Garside, P. (1995). 'Urban and economic development', in J. Stewart and G. Stoker (Eds) *Local Government in the 1990s*. London: MacMillan, pp. 166–187.

Hawkins, R. G. P. and Shaw, H. (2004). 'Sustainable development: A monument for eternity', *Engineering Sustainability*, 156(ESI), pp. 3–5.

Heal, G. M., Barbier, E. B., Boyle, K. J., Covich, A. P., Gloss, S. P., Hershner, C. H., Hoehn, J. P., Pringle, C. M., Polasky, S., Segerson, K. and Shrader-Frechette, K. (2005). *Valuing Ecosystem Services: Toward Better Environmental Decision Making*. Washington, DC: The National Academies Press.

Helliwell, D. R. (1969). 'Valuation of wildlife resources', *Regional Studies*, 3, pp. 41–49.

Hemphill, L., Berry, J. and McGreal, S. (2004). 'An indicator-based approach to measuring sustainable urban regeneration performance: Conceptual foundations and methodological framework', *Urban Studies*, 41, pp. 725–772.

Her Majesty's Stationery Office (HMSO). (1972). *Bains Report, The New Local Authorities: Management and Structure*. London: HMSO.

Her Majesty's Stationery Office (HMSO). (1995). *A Further Assessment of the Enterprise Zone Policy*. London: HMSO.

Higham, A., Fortune, C. and James, H. (2015). 'Life cycle costing: Evaluating its use in UK practice', *Structural Survey*, 33(1), pp. 73–87.

Higham, A. P., Barlow, C., Bichard, E. and Richards, A. (2017). 'Valuing sustainable change in the built environment: Using SuROI to appraise built environment projects', *Journal of Facilities Management* (in Press).

Higham, A. P., Fortune, C. J. and Boothman, J. C. (2016). 'Sustainability and investment appraisal for housing regeneration projects', *Structural Survey*, 34(2), pp. 150–167.

Hilditch, M. (2009). *Spot the Difference*. Available at: http://www.insidehousing.co.uk/story.aspx?storycode=6505317 (Accessed 01/02/2018).

Hills, J. (2007). *Ends and Means: The Future Roles of Social Housing in England*. Available at: eprints.lse.ac.uk/.../Ends_and_Means_The_future_roles_of_social_housing_in_England (Accessed 17/12/2017).

HMSO, Her Majesty's Government. (1994). *Sustainable Development. The UK Strategy*. London: HMSO.

H M Treasury. (1988). *Policy Evaluation: A Guide for Managers*. London: HMSO.

H M Treasury. (1995). *A Framework for the Evaluation of Regeneration Projects and Programmes*. London: HMSO.

H M Treasury. (2018). *The Green Book*. Available at: https://www.gov.uk/government/uploads/system/uploads/attachment_data/file/685903/The_Green_Book.pdf (Accessed 01/03/2018).

Ho, S. Y. (1998). *An Evaluation Agenda: From Value for Money to Local Decision Making*. Leeds: CUDEM Working Paper, Leeds Metropolitan University.

Ho, S. Y. (1999). 'Evaluating urban regeneration programmes in Britain: Exploring the potential of the realist approach', *Evaluation*, 5(4), pp. 422–438.

Ho, S. Y. (2003). *Evaluating British Urban Policy: Ideology, Conflict and Compromise*. Farnham: Ashgate.

Hogan, R. L. (2007). 'The historical development of program evaluation: Exploring the past and present', *Online Journal of Workforce Education and Development*, 2(4), Available at: https://pdfs.semanticscholar.org/ee2f/dbbe116a30ab7a79b19e1033a7cab434feec.pdf (Accessed 07/02/2018).

Holland, A. (2003). 'Sustainability', in D. Jamieson (Ed) *A Companion to Environmental Philosophy*. Oxford: Blackwell Publishing, pp. 390–401.

Holmberg, J. (Ed). (1994). *Policies for a Small Planet*. London: Earthscan.

Homes and Communities Agency (HCA). (2012). *The Regulatory Framework for Social Housing in England from April 2012*. London: HCA.

Hood, C. (1991). 'A public management for all seasons?' *Public Administration*, 69(1), pp. 3–19.

Hornberg, C. and Pauli, A. (2011). 'Substandard housing: The social dimension of environmental health', in J. O. Nriagu (Ed) *Encyclopedia of Environmental Health*. Volume 2. Burlington, MA: Elsevier, pp. 276–289.

Hoskins, K. (1968). 'The examination, disciplinary power and rational schooling', *History of Education*, 8, pp. 135–146.

House of Commons, Communities and Local Government Committee. (2009). *Housing and the Credit Crunch Third Report of Session 2008–09*. Available at: http://www.publications.parliament.uk/pa/cm200809/cmselect/cmcomloc/101/101.pdf (Accessed 01/02/2018).

House of Lords. (2015). *The Select Committee on National Policy for the Built Environment Inquiry on Built Environment*. Evidence Session Number 5.

Housing Associations' Charitable Trust (HACT). (2018). *Social Value Bank*. Available at: http://www.hact.org.uk/social-value-bank (Accessed 01/03/2018).

Howaldt, J., Butzin, A., Domanski, D. and Kaletka, C. (2014). *Theoretical Approaches to Social Innovation – A Critical Literature Review*. A deliverable of the project: 'Social Innovation: Driving Force of Social Change' (SI-DRIVE). Dortmund: Sozialforschungsstelle.

Huby, M. (1998). *Social Policy and the Environment*. Buckingham: Open University Press.

Imperial College London. (2007). *Project Stakeholder Analysis*. Available at: www3.imperial.ac.uk (Accessed 11/11/17).

Imrie, R. and Raco, M. (2003). *Urban Rennaissance? New Labour, Community and Urban Policy*. Bristol: The Policy Press.

Imrie, R. and Thomas, H. (1995). 'Changes in local governance and their implications for urban policy evaluation', in R. Hambleton and H. Thomas (Eds) *Urban Policy Evaluation*. London: Paul Chapman Publishing, pp. 123–138.

Imrie, R. and Thomas, H. (1999). *British Urban Policy: An Evaluation of the Urban Development Corporations*. (2nd Ed). London: Sage.

Inside Housing. (2014). *Sharp Rise in Turnover from Commercial Activity*. Available at: http://www.insidehousing.co.uk/exclusive-sharp-rise-in-turnover-from-commercial-activity/7006908.article (Accessed 01/12/2017).

International Federation of Arts Councils and Culture Agencies. (2006). *Arts And Culture in Regeneration. D'Art topics in Arts Policy, No 25*. Sydney: International Federation of Arts Councils and Culture Agencies.

Ireh, M. (2016). *Scientific Management Still Endures in Education*. Available at: https://files.eric.ed.gov/fulltext/ED566616.pdf (Accessed 01/02/2018).

IUCN/UNEP/WWF. (1991). *Caring for the Earth: A Strategy for Sustainable Living. International Union for the Conservation of Nature*. Switzerland: Gland.

IUCN/UNEP/WWF/FAO/UNESCO. (1980). *World Conservation Strategy: Living Resource Conservation for Sustainable Development*. Switzerland: IUCN, Gland.

Jacobs, D. E., Brown, M. J., Baeder, A., Sucosky, M. S., Margolis, S. and Hershovitz, J. (2010). 'A systematic review of housing interventions and health: Introduction, methods, and summary findings', *Journal of Public Health Management and Practice*, 16, pp. S5–S10.

Jeffrey, P. and Pounder, J. (2000). 'Physical and environmental aspects', in P. Roberts and H. Sykes (Eds) *Urban Regeneration: A Handbook*. London: Sage, pp. 86–108.

Jepsen, A. L. and Eskerod, P. (2009). 'Stakeholder analysis in projects: Challenges in using current guidelines in the real world', *International Journal of Project Management*, 27(4), pp. 335–343.

Johnson, G. and Scholes, K. (2002). *Exploring Corporate Strategy*. (6th Ed). Harlow: Pearson Education.

Johnson, R. (2013). 'Pervasive interactions: A purposive best evidence review with methodological observations on the impact of housing circumstances and housing interventions on adult mental health and well-being', *Housing, Care and Support*, 16(1), pp. 32–49.

Jones, P. and Evans, J. (2008). *Urban Regeneration in the UK*. London: Sage.

Jones, P. S. (2003). 'Urban regeneration's poisoned chalice: Is there an impasse in (community) participation-based policy?', *Urban Studies*, 40(3), pp. 581–601.

Jonker, J. and Foster, D. (2002). 'Stakeholder excellence? Framing the evolution and complexity of a stakeholder perspective of the firm', *Corporate Social Responsibility and Environmental Management*, 9, pp. 187–195.

Joyner, C. (2005). 'Rethinking international environmental regimes: What role for partnership coalitions?' *Journal of International Law and International Relations*, 1(1–2), pp. 89–119.

Julnes, G., Mark, M. M. and Henry, G. T. (1998). 'Promoting realism in evaluation: Realistic evaluation and the broader context', *Evaluation*, 4(4), pp. 483–504.

Kanter, R. M. (1999). 'From spare change to real change', *Harvard Business Review*, 77, pp. 122–132.

Karlsen, J. T. (2002). 'Project stakeholder management', *Engineering Management Journal*, 14(4), pp. 19–24.

Kazmierczak, A. E., Curwell, S. R. and Turner, J. C. (2009). 'Regeneration of large urban area: Assessment methods', *Proceedings of the Institute of Civil Engineer*, Issue ME2, pp. 117–124.

Keles, R. (2003). 'Urban regeneration in Istanbul'. *Paper presented to Priority Action Programme*, Regional Activity Centre, Split.

Kellaghan, T. and Stufflebeam, D. L. (Eds). (2003). *International Handbook of Educational Evaluation*. Dordrecht: Kluwer, pp. 701–720.

Kerzner, H. (2011). *Project Management Metrics, KPIs, and Dashboards: A Guide to Measuring and Monitoring Project Performance*. Hoboken, NJ: Wiley and Sons.

Kettl, D. (2002). *The Transformation of Governance: Public Administration for Twenty-First Century America*. Baltimore, MD: Johns Hopkins University Press.

Khakasa, E. (2009). *Capital Budgeting for Information Systems Investments – Evidence from Kenya*. Available at: http://www. ssrn.com/abstract=2026392 (Accessed 07/01/2018).

King, P. (2012). 'The ethics of affordable housing', in E. Chadwick (Ed) *Encyclopaedia of Applied Ethics*. (2nd Ed). San Diego, CA: Academic Press, pp. 72–78.

King, R. T. (1966). 'Wildlife and man', *New York Conservationist*, 20(6), pp. 8–11., also in Bailey, I. A., Elder, W. and McKinney, T. B. (1974). *Readings in Wildlife Conservation*. Washington, DC: The Wildlife Society.

Kleinhans, R. (2012). *Selling Houses to (Presumably) Low-Income Tenants Can Owner-occupation lead to Empowerment of Former Social Housing Tenants?* Delft: Delft University of Technology, OTB Research Institute for Housing, Urban and Mobility Studies, The Netherlands.

Krlev, G., Münscher, R. and Mülbert, K. (2013). *Social Return on Investment (SROI): State-of the Art and Perspectives: A Meta-Analysis of Practice in Social Return on Investment (SROI) Studies Published 2000–2012*. Available at: https://www.csi.uniheidelberg.de/downloads/CSI_SROI_Meta_Analysis_2013.pdf (Accessed 24/12/2017).

Lam, P. T. I., Chan, E. H. W., Chau, C. K. and Poon, C. S. (2011). 'A sustainable framework of "green" specification for construction in Hong Kong', *Journal of Facilities Management*, 9(1), pp. 16–33.

Langston, C. A. and Ding, G. K. C. (2001). *Sustainable Practices in the Built Environment*. (2nd Ed). Oxford: Butterworth Heinemann.

Lawless, P. (1989). *Britain's Inner Cities*. London: Paul Chapman Publishing Ltd.

Lawless, P. (2011). 'Understanding the scale and nature of outcome change in area-regeneration programmes: Evidence from the new deal for communities programme in England', *Environment and Planning* C, 29, pp. 520–532.

Lawless, P., Pearson, S., Wilson, I., Batty, E., Beatty, C. and Foden, M. (2010). *The New Deal for Communities Experience: A Final Assessment*. London: DCLG.

Lawson, L. and Kearns, A. (2010). 'Community engagement in regeneration: Are we getting the point?' *Journal of Housing and the Built Environment*, 25(1), pp. 19–36.

Lee, G. and Chan, E. (2007). 'The analytic hierarchy process (AHP) approach for assessment of urban renewal proposals', *Social Indicators Research*, 89(1), pp. 155–168.

Lee, G. and Chan, E. (2008). 'Indicators for evaluating environmental performance of the Hong Kong urban renewal projects', *Facilities*, 27, pp. 515–530.

Lehtonen, M. (2004). 'The environmental-socio interface of sustainable development: Capabilities, social capital, institutions', *Ecological Economics*, 49, pp. 199–214.

Lewis, S. and Jones, J. (1990). 'The use of output and performance measures in government departments', in M. Cave, M. Kogan and R. Smith (Eds) *Output and Performance Measurement in Government: The States of the Art*. London: Jessica Kingsley.

LGA (Local Government Association). (2000). *A Change of Scene: The Challenges of Tourism in Regeneration*. London: LGA/DCMS.

Lingane, A. and Olsen, S. (2004). 'Guidelines for social return on investment', *California Management Review*, 46, pp. 116–135.

Littig, B. and Griessler, E. (2005). 'Social sustainability: A catchword between political pragmatism and social theory', *International Journal of Sustainable Development*, 8, pp. 65–79.

Liu, X. (2010). 'Housing renewal policies, house prices and urban competitiveness', *Applied Geography*, 30(2), pp. 221–228.

Lombardi, R., Porter, L., Barber, A. and Rogers, C. (2011). 'Conceptualising sustainability in UK urban regeneration: A discursive formation', *Urban Studies*, 48(2), pp. 273–296.

Lumley, S. and Armstrong, P. (2004). 'Some of the nineteenth century origins of the sustainability concept', *Environment, Development and Sustainability*, 6, pp. 367–378.

MacArthur, J. (1997). 'Stakeholder analysis in project planning: Origins, applications and refinements of the method', *Project Appraisal*, 12(4), pp. 251–265.

MacLennon, D. (2007). *Better Futures for Social Housing in England*. Available at: https://www.jrf.org.uk/report/better-futures-social-housing-england (Accessed 05/08/2017).

Madaus, G. F. and Kellaghan, T. (1982). 'Trends in standards in Great Britain and Ireland', in G. Austin and H. Garber (Eds) *The Rise and Fall of National Test Scores*. New York: Academic Press, pp. 195–214.

Madaus, G. F., Stufflebeam, D. L. and Kellaghan, T. (2000). *Evaluation Models: Viewpoints on Educational and Human Services Evaluation*. (2nd Ed). Hingham, MA: Kluwer Academic Publishers.

Maddedu, M. (2013). 'Housing quality and the rescue of failed private housing schemes in England: A policy review', *The Journal of Housing and the Built Environment*, 28(3), pp. 567–578.

Madlener, R., Robledo, C., Muys, B., Hektor, B. and Domac, J. (2003). *A Sustainability Framework for Enhancing the Long-Term Success of LULUCF Projects*. Berlin: Springer.

Mah, A. (2012). 'Demolition for development: A critical analysis of official urban Imaginaries in past and present UK cities', *Journal of Historical Sociology*, 25(1), pp. 151–176.

Malena, C. (2004). *Strategic Partnership: Challenges and Best Practices in the Management of Governance of Multi-Stakeholder Partnerships Involving UN and Civil Society Actors*, Background Paper, Multi-Stakeholder Workshop on Partnerships and UN-Civil Society. New York: Relations, Pocantico.

Maliene, V., Howe, J. and Malys, N. (2008). 'Sustainable communities: Affordable housing and socio-economic relations', *Local Economy*, 23(4), pp. 267–276.

Maliene, V. and Malys, N. (2009). 'High-quality housing – A key issue in delivering sustainable communities', *Building and Environment*, 44(1), pp. 426–430.

Malpass, P. (2005). *Housing and the Welfare State: The Development of Housing Policy in Britain*. London: Palgrave MacMillan.

Maughan, C. (2012). *Monitoring and Evaluating Social Impacts in Australia*, CRC-REP Working Paper, CW003. Alice Springs: Ninti One Limited.

Mawih, K. (2015). 'A strategic framework to use payback period in evaluating the capital budgeting in energy and oil and gas sectors in Oman', *International Journal of Economics and Financial Issues*, 5(2), pp. 469–475.

McGraw, C. (2007). 'Housing and health', *Primary Health Care*, 17(4), p. 35.

McGregor, A. and McConnachie, M. (1995). 'Social exclusion, urban regeneration and economic re-integration', *Urban Studies*, 32(10), pp. 1587–1600.

McQuaid, R. W., Greig, M. and Lindsay, C. (2006). *Approaches to Evaluation in Community Regeneration: A Report to Communities Scotland*. Edinburgh: Employment Research Institute, Napier University, Edinburgh.

Mebratu, D. (1998). 'Sustainability and sustainable development: Historical and conceptual review', *Environmental Impact Assessment Review*, 18(6), pp. 493–520.

Millar, R. and Hall, K. (2013). 'Social Return on Investment (SROI) and performance measurement: The opportunities and barriers for social enterprises in health and social care', *Public Management Review*, 15(6), pp. 923–941.

Ministry of Housing, and Local Communities (MHLC). (2019). *English Housing Headline Report 2017–18*. London UK Government.

Moorhouse Consulting. (2007). *Beyond Conventional Stakeholder Management: Developing PRIME Intelligence on Complex Programmes*. Available at: www.moorhouseconsulting.com (Accessed 01/01/2018).

Morgan, P. and Taschereau, S. (1996). *Capacity and Institutional Assessment: Frameworks, Methods and Tools for Analysis*, Prepared for CIDA Policy Branch, Available at: http://www.acdicida.gc.ca/INET/IMAGES.NSF/vLUImages/CapacityDevelopment$-file/1996-06-02Tools:Wkshp6).pdf (Accessed 16/11/2017).

Moulaert, F., MacCallum, D., Mehmood, A. and Hamdouch, A. (2013). *The International Handbook on Social Innovation. Collective Action, Social Learning and Transdisciplinary Research*. Cheltenham: Elgar.

Murray-Webster, R. and Simon, P. (2006). 'Making sense of stakeholder mapping', *PM World Today*, 8(11), pp. 1–5.

Mushove, P. and Vogel, C. (2005). 'Heads or tails? Stakeholder analysis as a tool for conservation area management', *Global Environmental Change*, 15, pp. 184–198.

Musto, V. (2015). *A Guide to Social Return on Investment for Alcohol and Drug Treatment Commissioners*. London: Public Health England.

Naess, A. (1973). 'The shallow and the deep, long-range ecology movement: A summary', *Inquiry: An Interdisciplinary Journal of Philosophy and the Social Sciences*, 16(1), pp. 95–100.

Neumayer, E. (2003). *Weak versus Strong Sustainability: Exploring the Limits of Two Opposing Paradigms*. Northampton: Edward Elgar.

Neumayer, E. (2012). 'Human development and sustainability', *Journal of Human Development and Capabilities*, 13(4), pp. 561–579.

Newcombe, R. (2003). 'From client to project stakeholders: A stakeholder mapping approach', *Construction Management and Economics*, 21, pp. 841–848.

New Economics Foundation (NEF). (2004). *Social Return on Investment: Valuing What Matters*. London: New Economics Foundation.

New Philanthropy Capital (NPC). (2010). *Social Return on Investment*. London: New Philanthropy Capital.

NHF. (2014). *An Ambition to Deliver. Housing Associations Unbounded*. Available at: http://s3-eu west1.amazonaws.com/pub.housing.org.uk/An_ambition_to_deliver.pdf (Accessed 01/02/2018).

Nicholls, J., Lawlor, E., Neizert, E. and Goodspeed, T. (2012). *A Guide to Social Return on Investment*. Liverpool: SROI Network.

Nicholls, J., Mackenzie, S. and Somers, A. (2007). *Measuring Real Value*. London: New Economics Foundation.

Nutt, P. (2002). *Why Decisions Fail: Avoiding the Blunders and Traps That Lead to Debacles*. San Francisco, CA: Berrett-Koehler Publishers.

Oatley, N. (1989). 'Evaluation and urban development corporations', *Planning Practice and Research*, 4, pp. 6–12.

Odum, E. P. and Odum, H. T. (1972). 'Natural areas as necessary components of man's total environment', in *Transactions of the 37th North American Wildlife and Natural Resources Conference, March 12–15, 1972*. Washington, DC: Wildlife Management Institute, vol. 37, pp. 178–189.

Office for National Statistics (ONS). (2016). *Statistical Bulletin. The UK National Balance Sheet: 2016 Estimates*. London: ONS.

Office for National Statistics (ONS). (2018). *Office for National Statistics*. Available at: https://www.ons.gov.uk (Accessed 01/03/2018).

Office of Government Commerce UK. (2003). *Managing Successful Programmes*. London: The Stationary Office.

Office of the Deputy Prime Minister (ODPM). (2001). *Towns & Cities: Partners in Urban Renaissance. Breaking down the Barriers*. London: ODPM.

Office of the Deputy Prime Minister (ODPM). (2003). *Sustainable Communities: Building for the Future*. Available at: www.odpm.gov.uk/communities/plan/main/index.htm (Accessed 07/10/2017).

Office of the Deputy Prime Minister (ODPM). (2004). *Assessing the Impacts of Spatial Interventions Regeneration, Renewal and Regional Development 'The 3Rs guidance'*. London: ODPM.

Office of the Deputy Prime Minister (ODPM). (2005a). *Local Strategic Partnerships: A Consultation Paper*. London: Crown Copyright.

Office of the Deputy Prime Minister (ODPM). (2005b). *National Evaluation of Local Strategic Partnerships – Issues Paper: Local Strategic Partnerships, Multi-Level Governance and Economic Development*. London: ODPM.

Olander, S. (2006). *External Stakeholder Management*. Unpublished PhD Thesis, Lund University, Sweden.

Olsson, S. and Nicholls, J. (2005). *A Framework for Approaches to SROI Analysis*. Available at:https://ccednet-rcdec.ca/sites/ccednet-rcdec.ca/files/ccednet/pdfs/2005-050624_SROI_Framework.pdf (Accessed 09/02/2018).

Organisation for Economic Co-Operation and Development (OECD). (2000). *Urban Renaissance: Belfast's Lessons for Policy and Partnership*. Paris: OECD.

Organisation for Economic Co-Operation and Development (OECD). (2004). *Local Economic and Employment Development: How to Assess What Works among Programmes and Policies*. Paris: OECD Publications.

O'Riordan, T., Cameron, J. and Jordan, A. (2001). *Reinterpreting the Precautionary Principle*. London: Cameron May.

Overseas Development Administration (ODA). (1995). *Guidance Note on How to Do a Stakeholder Analysis of Aid Projects and Programmes*. London: Overseas Development Administration.

Oxford English Dictionary (OED). (2017). Available at: http://www.oed.com (Accessed 02/07/2017).

Paddison, R. (2012). 'Housing and neighbourhood quality: Urban regeneration', in *International Encyclopaedia of Housing and Home*. London: Elsevier, pp. 288–293.

Pajunen, K. (2006). 'Stakeholder influences in organizational survival', *Journal of Management Studies*, 43(6), pp. 1261–1288.

Parkin, S. (2000). 'Sustainable development: The concept and the practical challenge', *Proceedings of the Institution of Civil Engineers – Civil Engineering*, 138(6), pp. 3–8.

Pearce, D. and Atkinson, G. (1992). *Are National Economies Sustainable?: Measuring Sustainable Development*, CSERGE Discussion Paper GEC 92-11. London: University College, London.

Pearce, D., Markandia, A. and Barbier, E. (1989). *Blueprint for a Green Economy*. London: Earthscan.

Pearce, D. W. (1993). *Economic Values and the Natural World*. London: Earthscan.

Peattie, K. and Morley, A. (2008). *Social Enterprises: Diversity and Dynamics, Contexts and Contributions, a Research Monograph*. Cardiff: ESRC Centre for Business Relationships.

Pepper, D. (1996). *Modern Environmentalism: An Introduction*. London: Routledge.

Pike, R. H. (1985). *Disenchantment with DCF Promotes IRR*. Certified Accountant, *The CPA Journal*, July, pp. 14–17.

Pitt, M., Tucker, M., Riley, M. and Longden, J. (2009). 'Towards sustainable construction: promotion and best practices', *Construction Innovation*, 9(2), pp. 201–224.

Places for People. (2017). *Creating Aspirational Homes and Inspirational Places*. Available at: https://www.placesforpeople.co.uk/media/1055/annual-report-2017.pdf (Accessed 1 February 2018).

Pollitt, C. (1993). 'Occasional excursions: A brief history of policy evaluation in the UK', *Parliamentary Affairs*, 46(3), pp. 353–362.

Polonsky, M. and Grau, S. L. (2011). 'Assessing the social impact of charitable organizations – Four alternative approaches', *International Journal of Nonprofit and Voluntary Sector Marketing*, 16, pp. 195–211.

Pope, J., Annandale, D. and Morrison-Saunders, A. (2004). 'Conceptualising sustainability assessment', *Environmental Impact Assessment Review*, 24(6), pp. 595–616.

Potter, J. and Moore, B. (2000). 'UK Enterprise Zones and the attraction of inward investment', *Urban Studies*, 37, pp. 1279–1312.

Power, A. (2008). 'Does demolition or refurbishment of old and inefficient homes help to increase our environmental, social and economic viability?' *Energy Policy*, 36(12), pp. 4487–4501.

Priemus, H., Dieleman, F. and Clapham, D. (1999). 'Current developments in social housing management', *Journal of Housing and the Built Environment*, 4(3), pp. 211–223.

Priemus, H. and Ten Heuvelhof, E. (2005). 'The long way to sustainable housing areas', *Environment and Planning B: Planning and Design*, 32, pp. 1–3.

Project Management Institute. (2004). *A Guide to the Project Management Body of Knowledge*. Newtown Square, PA: Project Management Institute.

Raphael, D. (2000). 'The question of evidence in health promotion', *Health Promotion International*, 15, pp. 355–367.

Rapley, M. (2001). *Quality of Life Research*. London: Sage.

Rauschmayer, F. and Wittmer, H. (2006). 'Evaluating deliberative and analytical methods for the resolution of environmental conflicts', *Land Use Policy*, 23, pp. 108–122.

Reed, M. S. (2008). 'Stakeholder participation for environmental management: A literature review', *Biological Conservation*, 141, pp. 2417–2431.

Reed, M. S., Graves, A., Dandy, N., Posthumus, H., Hubacek, K. and Morris, J. (2009). 'Who's in and why? A typology of stakeholder analysis methods for natural resource management', *Journal of Environmental Management*, 90(5), pp. 1933–1949.

Rhodes, J., Tyler, P. and Brennan, A. (2007). *Evaluation of the Single Regeneration Budget: Final Evaluation*. Cambridge: Department of Land Economy, University of Cambridge.

Rhodes, R. A. W. (2000). 'Governance and public administration', in J. Pierre (Ed) *Debating Governance. Authority, Steering, and Democracy*. Oxford: Oxford University Press, pp. 55–90.

Rietbergen-McCracken, J. and Narayan, D. (1998). *Participation and Social Assessment: Tools and Techniques*. Washington, DC: World Bank. Available at: http://documents.worldbank.org/curated/en/673361468742834292/pdf/multi0page.pdf (Accessed 09/02/2018).

Roberts, P. (2000). *The New Territorial Governance: Planning, Developing and Managing the United Kingdom in an Era of Devolution*. London: Town and Country Planning Association.

Roberts, P. (2003). 'Sustainable development and social justice: Spatial priorities and mechanisms for delivery', *Sociological Inquiry*, 73, pp. 228–244.

Roberts, P. (2006). 'Evaluating regional sustainable development: Approaches, methods and the politics of analysis', *Journal of Environmental Planning and Management*, 49(4), pp. 515–532.

Roberts, P. and Benneworth, P. (2002). 'Devolution sustainability and local economic development: Impacts on local autonomy, policy-making and economic development outcomes', *Local Economy*, 17, pp. 239–252.

Roberts, P. and Sykes, H. (2000). *Urban Regeneration: A Handbook*. London: Sage.

Robson, B., Bradford, M. and Parkinson, M. (1994). *Assessing the Impact of Urban Policy*. London: HMSO.

Rodriguez, S. I., Roman, M. S. and Sturhahn, S. C. (2002). *Sustainable Assessment and Report for the University of Michigan's Ann Arbor Campus*. Ann Arbor: University of Michigan.

Rohe, W. and Stegman, M. (1994). 'The impact of home ownership on the social and political involvement of low income people', *Journal of the American Planning Association*, 60(2), pp. 173–184.

Roloff, J. (2008). 'Learning from multi-stakeholder networks: Issue-focussed stakeholder management', *Journal of Business Ethics*, 82, pp. 233–250.

Rossi, P. H., Lipsey, M. W. and Freeman, H. E. (2004). *Evaluation: A Systematic Approach*. (7th Ed). Thousand Oaks, CA: Sage.

Rotheroe, N. and Richards, A. (2007). 'Social return on investment and social enterprise: Transparent accountability for sustainable development', *Social Enterprise Journal*, 3(1), pp. 31–48.

Roult, R. and Lefebvre, S. (2013). 'Stadiums, public spaces and mega-events: Cultural and sports facilities as catalysts for urban regeneration and development', in M. L. Leary and J. McCarthy (Eds) *The Routledge Companion to Urban Regeneration*. London and New York: Routledge, pp. 548–557.

Rowlands, R. (2010). *Co-operative and Mutual Housing in the Social Rented Sector*. Third Sector Research Centre, Working Paper 17. Birmingham: Third Sector Research Sector.

Rowlinson, S. and Cheung, Y. K. F. (2008). 'Stakeholder management through empowerment: Modelling project success', *Construction Management and Economics*, 26(6), pp. 611–623.

Royal Institute of British Architects (RIBA). (2017). *RIBA Manifesto 2017: Building Global Britain*. Available at: https://www.architecture.com/-/media/gathercontent/building-global-britain/additional-documents/11056ribamanifesto2017pdf (Accessed 01/12/2017).

Royal Institution of Chartered Surveyors (RICS). (2012). *Financial Viability in Planning, RICS Guidance Note*. Coventry: Royal Institution of Chartered Surveyors.

Royal Institution of Chartered Surveyors (RICS). (2014). *RICS Valuation – Professional Standards*. London: Royal Institution of Chartered Surveyors.

Royal Institution of Chartered Surveyors (RICS). (2017). *RICS Valuation – Professional Standards*. London: Royal Institution of Chartered Surveyors.

Russell, G., Johnston, T. and Pritchard, J. (2000). *Final Evaluation of City Challenge; KPMG Consulting Commissioned by Department of the Environment Transport and the Regions*. London: DETR.

Russell, H., Dawson, J., Garside, P. and Parkinson, M. (1996). *City Challenge: Interim National Evaluation*. London: The Stationery Office.

Ryan, P. A. and Ryan, G. P. (2002). 'Capital budgeting practices of the fortune 1000: How have things changed?', *Journal of Business and Management*, 8(4), pp. 355–364.

Rydin, Y., Bleahu, A., Davies, M., Davila, J. D., Friel, S. and De Grandis, G. (2012). 'Shaping cities for health: Complexity and the planning of urban environments in the 21st century', *The Lancet*, 379(9831), pp. 2079–2108.

Schmeer, K. (1999). *Guidelines for Conducting a Stakeholder Analysis*. Bethesda, MD: Partnerships for Health Reform, Abt Associates Inc.

Scholten, P., Nicholls, J., Olsen, S. and Galimidi, B. (2006). *Social Return on Investment: A Guide to SROI Analysis*. Amsterdam: Lenthe Publishers.

Schwartz, A. (2011). 'The credit crunch and subsidized low-income housing: The UK and US experience compared', *Journal of Housing and the Built Environment*, 26(3), pp. 353–374.

Scottish Government. (2011). *Achieving a Sustainable Future. Regeneration Strategy*. Edinburgh: Scottish Government.

Scriven, M. (1996). 'The theory behind practical evaluation', *Evaluation*, 2(4), pp. 393–404.

Segelod, E. (1995). *Resource Allocation in Divisionalized Groups*. Avebury: Ashgate.

Seitanidi, M. M. and Crane, A. (2009). 'Implementing CSR through partnerships: Understanding the selection, design and institutionalisation of non-profit business partnerships', *Journal of Business Ethics*, 85, pp. 413–429.

Semper, D. and Fuertes, I. (2015). 'Social return and financing of urban regeneration policies', in *Conference: 51st European Congress of the Regional Association International*. Barcelona, Spain.

Shaw, K. (1995). 'Assessing the performance of urban development corporations: The reliability of the official government output measures', *Planning Practice and Research*, 19(3), pp. 289–97.

Shelton, D., Cork, S., Binning, C., Parry, R., Hairsine, P. and Vertessy, R. (2001). 'Application of an ecosystem services inventory approach to the Goulburn Broken Catchment'. *Third Australian Stream Management Conference, Brisbane, August 27–29, Cooperative Research Centre for Catchment Hydrology*, pp. 157–162.

Shinoda, T. (2010). 'Capital budgeting management practices in Japan', *Economic Journal of Hokkaido University*, 39, pp. 39–50.

Shirey, M. R. (2012). 'Stakeholder analysis and mapping as targeted communication strategy', *Journal of Nursing Administration*, 42(9), pp. 399–403.

Sloan, P. (2009). 'Redefining stakeholder engagement: From control to collaboration', *Journal of Corporate Citizenship*, 36, pp. 25–40.

Smith, N. L. (1987). *Army Ordnance and American System of Manufacturing*. Cambridge, MA: MIT Press.

Smith, R. (2006). 'Housing stock transfer: Investing in renewal as a tool for sustainable regeneration', *Housing Studies*, 21, pp. 269–282.

Smudde, P. M. and Courtright, J. L. (2011). 'A holistic approach to stakeholder management: A rhetorical foundation', *Public Relations Review*, 37(2), pp. 137–144.

Social Housing. (2016). *Non-core Turnover Increases 18.1 per cent in a Single Year*. Available at: https://www.socialhousing.co.uk/insight/insight/non-core-turnover-increases-181-per-cent-in-a-single-year-26144 (Accessed 01/11/2017).

Social Value UK. (2018). *SROI Impact Map Template*. Available at: http://www.socialvalueuk.org/resource/blank-value-map (Accessed 10/02/2018).

Solow, R. M. (1993). 'Sustainability: An economist's perspective', in R. Dorfman and N. S. Dorfman (Eds) *Economics of the Environment: Selected Readings*. New York: W. W. Norton, pp. 179–187.

Somerville, M., Mackenzie, I., Owen, P. and Miles, D. (2000). 'Housing and health: Does installing heating in their homes improve the health of children with asthma?', *Public Health*, 114(6), pp. 434–439.

SROI Network. (2011). *SROI Network Newsletter*. Available at: http://www.socialvalueuk.org/app/uploads/2016/03/SROI%20Network%20Fourth%20Quarterly%20Newsletter%20August%202011.pdf (Accessed 03/01/2018).

Stanford University. (1994). *Stanford University Press Release*. Available at: https://web.stanford.edu/dept/news/pr/94/940228Arc4425.html (Accessed 02/07/2017).

Steele, A. (2012). *A New Role for Housing Associations*. Available at: http://ggsrv-cold.standrews.ac.uk/chr/Uploads/Edit/file/A%20New%20Role%20for%20Housing%20Associations_Alister%20Steele.pdf (Accessed 01/12/2017).

Stewart, J. (2005). 'A review of UK housing policy: Ideology and public health', *Public Health*, 119, pp. 525–534.

Stewart, J. and Rhoden, M. (2006). 'Children, housing and health', *International Journal of Sociology and Social Policy*, 26(7/8), pp. 326–341.

Stoddart, H. (2011). *A Pocket Guide to Sustainable Development Governance*. London: Stakeholder Forum.

Stouffer, S. A., Lumsdaine, A. A., Lumsdaine, M. H., Williams, R. M., Smith, M. B., Janis, I. L., Star, S. A. and Cottrell, L. S. (1949). *The American Soldier: Combat and Its Aftermath, Vol II*. Princeton, NJ: Princeton University Press.

Stringham, E. (2001). 'Kaldor-Hicks efficiency and the problem of central planning', *The Quarterly Journal of Austrian Economics*, 4(2), pp. 41–50.

Strong, M. F. (1972). 'The Stockholm conference – Where science and politics meet', *Ambio*, 1, pp. 73–78.

Stufflebeam, D. L. and Shinkfield, A. J. (2007). *Evaluation Theory, Models and Applications*. San Francisco, CA: Jossey-Bass.

Sullivan, L. (2012) *RIBA Guide to Sustainability in Practice*. https://issuu.com/ribacomms/docs/ribaguidetosustainabilityinpractice2.

Tallon, A. (2009). *Urban Regeneration in the UK*. London: Routledge.

Taylor, D. and Balloch, S. (Eds). (2005). *The Politics of Evaluation*. Bristol: The Policy Press.

Temple, N., Wigglesworth, C. and Smith, C. (2014). *Communities Count: The Four Steps to Unlocking Social Value*. London: Social Enterprise UK.

Ten Brink, P., Berghofer, A., Neuville, A., Schroter-Schlaack, C., Vakrou, A., White, S. and Wittmer, H. (2011). 'Chapter 10: Responding to the value of nature', in P. Ten Brink (Ed) *The Economics of Ecosystems and Biodiversity for National and International Policy Makers*. London: Earthscan.

The Economics of Ecosystems and Biodiversity (TEEB). (2011). *The Economics of Ecosystems and Biodiversity in National and International Policy Making*. P. Ten Brink (Ed). London: Earthscan.

The University of Edinburgh. (2014). *SURegen Urban Regeneration Workbench.* https://www.epcc.ed.ac.uk/projects-portfolio/suregen-urban-regeneration-workbench.

The Work Foundation. (2012). *People or Place? Urban Policy in the Age of Austerity.* Available at:http://www.theworkfoundation.com/downloadpublication/report/321_updated_people%20or%20place_urban%20policy%20in%20the%20age%20of%20austerity.pdf (Accessed 04/07/2017).

Thomas, D. H. L. and Adams, W. M. (1997). 'Space, Time and Sustainability in the Hadejia-Jama' Are Wetlands and the Komodugu Yobe Basin, Nigeria', *Transactions of the Institute of British Geographers*, 22(4), pp. 430–449.

Thomson, C. S., El-Haram, M. A. and Hardcastle, C. (2009). 'Managing knowledge of urban sustainability assessment', *Engineering Sustainability*, 162(ES1), pp. 35–43.

Thomson, H., Morrison, D. and Petticrew, M. (2007). 'The health impacts of housing-led regeneration: A prospective controlled study', *Journal of Epidemiology and Community Health*, 61, pp. 211–14.

Thomson, H. and Thomas, S. (2015). 'Developing empirically supported theories of change for housing investment and health', *Social Science and Medicine*, 124, pp. 205–214.

Thomson, H., Thomas, S., Sellstrom, E. and Petticrew, M. (2013). *Housing Improvements for Health and Associated Socioeconomic Outcomes (Review).* Available at: www.ncbi.nlm.nih.gov/pubmed/23450585 (Accessed 09/09/2017).

Thornley, A. (1993). *Urban Planning under Thatcherism: The Challenge of the Market.* (2nd Ed). London: Routledge.

Tosics, I. (2004). 'European urban development: Sustainability and the role of housing', *Journal of Housing and the Built Environment*, 19, pp. 67–90.

Trotter, L., Vine, J., Leach, M. and Fujiwara, D. (2014). *Measuring the Social Impact of Community Investment: A Guide to Using the Wellbeing Valuation Approach.* London: HACT.

Truex, R. and Soreide, T. (2010). 'Why multi-stakeholder groups succeed and fail': Policy Research Working Paper', *The World Bank Sustainable Development Network, Finance, Economics and Urban Development Unit World Bank.*

Tsenkova, S. (Ed). (2003). *Urban Regeneration: Learning from the British Experience.* Calgary: Faculty of Environmental Design, University of Calgary.

Tuan, M. (2008). *Measuring and /or Estimating Social Value Creation: Insights into Eight Integrated Cost Approaches.* Prepared for Bill and Melinda Gates Foundation Impact Planning and Improvement.

Turner, R. K. (Ed). (1993). *Sustainable Environmental Economics and Management: Principles and Practice.* London: Bellhaven Press.

Turok, I. (1989). 'Evaluation and understanding in local economic policy', *Urban Studies*, 26(6), pp. 587–606.

Turok, I. (2005). 'Urban regeneration: What can be done and what should be avoided?', in *Istanbul 2004 International Urban Regeneration Symposium: Workshop of Kucukcekmece District.* Istanbul: Kucukcekmece Municipality Publication, pp. 57–62.

Tyler, P., Warnock, C., Provins, A. and Lanz, B. (2013). 'Valuing the benefits of urban regeneration', *Urban Studies*, 50(1), pp. 169–190.

UK Government. (1999). *Department of the Environment, Transport and Regions, Local Evaluation for Regeneration Partnerships – Good Practice Guide.* London: DETR.

UK Government. (2010). *Measuring Social Value – How Five Social Enterprises Did It. Social Enterprise Unit.* London: Crown Copyright. Available at: https://www.gov.uk/government/uploads/system/uploads/attachment_data/file/215895/dh_122354.pdf (Accessed 09/02/2018).

UK Government. (2017). *Fixing Our Broken Housing Market. Housing White Paper.* UK Government: Crown Copyright. Available at: https://www.gov.uk/government/collections/housing-white-paper (Accessed 09/03/2018).

UK Parliament. (2003). *The Effectiveness of Government Regeneration Initiatives. Select Committee on Office of the Deputy Prime Minister: Housing, Planning, Local Government and the Regions, Seventh Report.* London: The United Kingdom Parliament.

UNEP. (2011). *Green Economy Report. Pathways to Sustainable Development and Poverty Eradication A Synthesis for Policy Makers.* Available at: https://sustainabledevelopment.un.org/content/documents/126GER_synthesis_en.pdf (Accessed 03/03/2018).

UN-Habitat. (2011). *UN-Habitat Annual Report 2010.* UN-Habitat: United Nations Human Settlements Programme.

United Nations (UN). (2012a). *The Future We Want, Outcome Document of the United Nations Conference on Sustainable Development, Rio de Janeiro, 20–22 June.* Available at: http://www.un.org/disabilities/documents/rio20_outcome_document_complete.pdf (Accessed 01/02/2018).

United Nations (UN). (2012b). World *Urbanization Prospects, the 2011 Revision.* New York: Department of Economic and Social Affairs Population Division, pp. 1–50.

United Nations (UN). (2015). *Report of the Inter-Agency and Expert Group on Sustainable Development Goal Indicators. Economic and Social Council 17 December 2015.* Available at: http://unstats.un.org/unsd/statcom/47th-session/documents/2016-2-IAEGSDGs-E.pdf (Accessed 01/02/2018).

United Nations (UN). (2016). *Regional Report to HABITAT III. HABITAT III Regional Report on Housing and Urban Development for the UNECE region. Towards a City Focused, People Centred and Integrated Approach to the New Urban Agenda.* Available at: https://orca.cf.ac.uk/96624/1/HABITAT_III_Regional_Report_for_UNECE_Region_-_draft_9.0_08_july_submission_with_foreward.pdf (Accessed 01/02/2018).

United Nations (UN). (2017). *United Nations New Urban Agenda.* Available at: http://habitat3.org/wp-content/uploads/NUA-English.pdf (Accessed 03/03/2018).

Urban Task Force. (1999). *Towards an Urban Renaissance*, Final Report. London: DETR.

Van Bueren, E. and de Jong, J. (2007). 'Establishing sustainability: Policy successes and failures', *Building Research & Information*, 35(5), pp. 543–556.

Vardakoulias, O. (2013). Social CBA and SROI: Economics in the policymaking 4. Published by NEF (The New Economics Foundation). Available at: www.neweconomics.org (Accessed 29/11/2017).

Walker, D. H. T., Bourne, L. and Rowlinson, S. (2008). *Stakeholders and the Supply Chain. Procurement Systems – A Cross Industry Project Management Perspective.* Abingdon: Taylor and Francis.

Watson, K. J., Evans, J., Karvonen, A. and Whitley, T. (2016). 'Capturing the social value of buildings; The promise of social return on investment', *Building and Environment*, 103, pp. 289–301.

Watson, K. J. and Whitley, T. (2016). 'Applying Social Return on Investment (SROI) to the built environment', *Building Research and Information.* Available at: http://www.tandfonline.com/doi/full/10.1080/09613218.2016.1223486 (Accessed 01/12/2017).

Weible, C. M. (2006). 'An advocacy coalition framework approach to stakeholder analysis: Understanding the political context of California marine protected area policy', *Journal of Public Administration Research and Theory*, 17, pp. 95–117.

Whitehead, C. (2004). *The Economic Framework for Housing, Paper 4 of CABE and RIBA, Housing Futures 2024.* London: CABE and RIBA.

Williams, T. (2002). 'Keeping the poor in their place', *Regeneration & Renewal*, 16.

Williams, T. (2007). 'Regeneration is as easy as ABC', *Regeneration & Renewal*, 40.

Wilson, M. A. and Carpenter, S. R. (1999). 'Economic valuation of freshwater ecosystem services in the United States 1971–1997', *Ecological Applications*, 9(3), pp. 772–783.

Winston, N. (2009). 'Urban regeneration for sustainable development: The role of sustainable housing?', *European Planning Studies*, 17(12), pp. 1781–1796.

Winston, N. (2010). 'Regeneration for sustainable communities? Barriers to implementing sustainable housing in urban areas', *Sustainable Development*, 18(6), pp. 319–330.

Wood, C. and Leighton, D. (2010). *Measuring Social Value: The Gap between Policy and Practice*. London: Demos.

World Commission on Environment and Development (WCED). (1987). *Our Common Future*. Oxford: Oxford University Press.

World Health Organisation (WHO). (1998). *Health Promotion Glossary*. Geneva: WHO.

World Health Organization (WHO). (2009). *Stakeholder Analysis*. Available at: www.who.int/hac/techguidance/.../stakeholder%20analysis%20ppt.pdf (Accessed 09/11/2017).

WWF. (2003). *Building Sustainably: How to Plan and Construct New Housing for the 21st Century*. London: WWF.

Xing, Y., Horner, R., El-Haram, M. and Bebbington, J. (2009). 'A framework model for assessing sustainability impacts of urban development', *Accounting Forum*, 33, pp. 209–224.

Xue, J. (2012). 'Potentials for decoupling housing-related environmental impacts from economic growth', *Environmental Development*, 4, pp. 18–35.

Yang, J., Shen, G., Ho, M., Drew, D. and Chan, A. (2009). 'Exploring critical success factors for stakeholder management in construction projects', *Journal of Civil Engineering and Management*, 15(4), pp. 337–348.

Yang, R. J. (2014). 'An Investigation of stakeholder analysis in urban development projects: Empirical or rationalistic perspectives', *International Journal of Project Management*, 32, pp. 838–849.

Yau, Y. and Chan, H. L. (2008). 'To rehabilitate or redevelop? A study of the decision criteria for urban regeneration projects', *Journal of Place Management and Development*, 1(3), pp. 272–291.

Young, O., Berkhout, F., Gallopin, G., Janssen, M., Ostrom, E. and Vanderleeuw, S. (2006). 'The globalization of socio-ecological systems: An agenda for scientific research', *Global Environmental Change*, 16(3), pp. 304–316.

Young, T. L. (2006). *Successful Project Management*. (2nd Ed). London: Kogan Page.

Zappala, G. and Lyons, M. (2009). *Recent Approaches to Measuring Social Impact on the Third Sector: An Overview*, Background Paper No. 5, Centre for Social Impact, University of New South Wales.

Zari, M. P. (2012). 'Ecosystem services analysis for the design of regenerative built environments', *Building Research & Information*, 40(1), pp. 54–64.

Zheng, H. W., Shen, G. Q. and Wang, H. (2014). 'A review of recent studies on sustainable urban renewal', *Habitat International*, 41, pp. 272–279.

Zuberi, D. and Taylor, A. (2013). 'Urban renewal in Vancouver, Canada', in M. E. Leary and J. McCarthy (Eds) *The Routledge Companion to Urban Regeneration* (Routledge Companions). London: Routledge; New York: Taylor & Francis Group.

Index

Note: **Bold** page numbers refer to tables; *italic* page numbers refer to figures.

Made in the USA
Monee, IL
13 September 2026

62978983R10092